Water and Salt Balance of Great Salt Lake, Utah, and Simulation of Water and Salt Movement Through the Causeway

By STEVEN R. WOLD, BLAKEMORE E. THOMAS, and KIDD M. WADDELL

Prepared in cooperation with the
Utah Department of Natural Resources,
Division of State Lands and Forestry

U.S. GEOLOGICAL SURVEY WATER-SUPPLY PAPER 2450

U.S. DEPARTMENT OF THE INTERIOR
BRUCE BABBITT, Secretary

U.S. GEOLOGICAL SURVEY
Gordon P. Eaton, Director

Any use of trade names in this publication is for descriptive purposes only and does not imply endorsement by the U.S. Government.

For sale by the
U.S. Geological Survey
Branch of Information Services
Box 25286 Federal Center
Denver, CO 80225

Library of Congress Cataloging in Publication Data

Wold, Steven R.
Water and salt balance of Great Salt Lake, Utah, and simulation of water and salt movement through the causeway / by Steven R. Wold, Blakemore E. Thomas, and Kidd M. Waddell.
p. cm.—(U.S. Geological Survey water-supply paper ; 2450)
"Prepared in cooperation with the Utah Department of Natural Resources, Division of State Lands and Forestry."
Includes bibliographical references (p. -)
Supt. of Docs. no.: I 19.13:W2450
1. Lake circulation—Utah—Great Salt Lake. 2. Salinity—Utah—Great Salt Lake. I. Thomas, Blakemore E. II. Waddell, K. M. (Kidd M.), 1937- . III. Utah. Division of State Lands and Forestry. IV. Title. V. Series
GB1625.UBW65 1997 96-17521
551.48'2'0979242—dc20 CIP
ISBN 0–607–86822–8

CONTENTS

FIGURES

FIGURES—Continued

TABLES

CONVERSION FACTORS, VERTICAL DATUM, AND ABBREVIATED WATER-QUALITY UNITS

Multiply	By	To obtain
acre	0.4047	square hectometer
	4,047	square meter
acre-foot (acre-ft)	0.001233	cubic hectometer
	1,233	cubic meter
acre-foot per day [(acre-ft)/d]	0.014276	cubic meter per second
acre-foot per year [(acre-ft)/yr]	0.001233	cubic hectometer per year
cubic foot per second (ft^3/s)	0.02832	cubic meter per second
foot (ft)	0.3048	meter
foot per second (ft/s)	0.3048	meter per second
inch (in.)	25.4	millimeter
mile (mi)	1.609	kilometer
square foot (ft^2)	0.092903	square meter
ton (short)	0.9072	metric ton or megagram
square mile (mi^2)	2.589	square kilometer

The unit cubic foot per second (ft^3/s) is used in this report and also can be expressed as 1 ft^3/s = 1.9835 acre-ft/d.

Sea level: In this report, "sea level" refers to the National Geodetic Vertical Datum of 1929—a geodetic datum derived from a general adjustment of the first-order level nets of the United States and Canada, formerly called Sea Level Datum of 1929.

Chemical concentration is reported only in metric units. Chemical concentration is reported in grams per milliliter (g/mL) or grams per liter (g/L). Grams per liter is a unit expressing the solute per unit volume (liter) of water.

Water and Salt Balance of Great Salt Lake, Utah, and Simulation of Water and Salt Movement Through the Causeway

By Steven R. Wold, Blakemore E. Thomas, *and* Kidd M. Waddell

Abstract

The water and salt balance of Great Salt Lake primarily depends on the amount of inflow from tributary streams and the conveyance properties of a causeway constructed during 1957–59 that divides the lake into the south and north parts. The conveyance properties of the causeway originally included two culverts, each 15 feet wide, and the permeable rock-fill material.

During 1980–86, the salt balance changed as a result of record high inflow that averaged 4,627,000 acre-feet annually and modifications made to the conveyance properties of the causeway that included opening a 300-foot-wide breach. In this study, a model developed in 1973 by Waddell and Bolke to simulate the water and salt balance of the lake was revised to accommodate the high water-surface altitude and modifications made to the causeway. This study, done by the U.S. Geological Survey in cooperation with the Utah Department of Natural Resources, Division of State Lands and Forestry, updates the model with monitoring data collected during 1980–86. This report describes the calibration of the model and presents the results of simulations for three hypothetical 10-year periods.

During January 1, 1980, to July 31, 1984, a net load of 0.5 billion tons of dissolved salt flowed from the south to the north part of the lake primarily as a result of record inflows. From August 1, 1984, when the breach was opened, to December 31, 1986, a net load of 0.3 billion tons of dissolved salt flowed from the north to the south part of the lake primarily as a result of the breach.

For simulated inflow rates during a hypothetical 10-year period resulting in the water-surface altitude decreasing from about 4,200 to 4,192 feet, there was a net movement of about 1.0 billion tons of dissolved salt from the south to the north part, and about 1.7 billion tons of salt precipitated in the north part. For simulated inflow rates during a hypothetical 10-year period resulting in a rise in water-surface altitude from about 4,200 to 4,212 feet, there was a net movement of about 0.2 billion tons of dissolved salt from the south to the north part and no salt was precipitated in the north part of the lake.

INTRODUCTION

Prior to construction of a railroad causeway during 1957–59, the hydrologic characteristics of Great Salt Lake, Utah, were typical of a closed lake having no outlet to the sea. After completion of the causeway (figs. 1 and 2) in 1959, the water and salt balance of the lake changed. The causeway divides the lake into a south and a north part (fig. 3). Slightly more than one-third of the surface area of the lake is north of the causeway. The formerly free movement of brine within the lake is now restricted by the causeway. Because more than 95 percent of freshwater surface inflow enters the lake south of the causeway, the causeway has interrupted the circulation and caused substantial changes to the hydrology and chemistry of the lake.

Waddell and Bolke (1973) described the effects of the causeway on the water and salt balance of Great Salt Lake and developed a model to simulate the effects

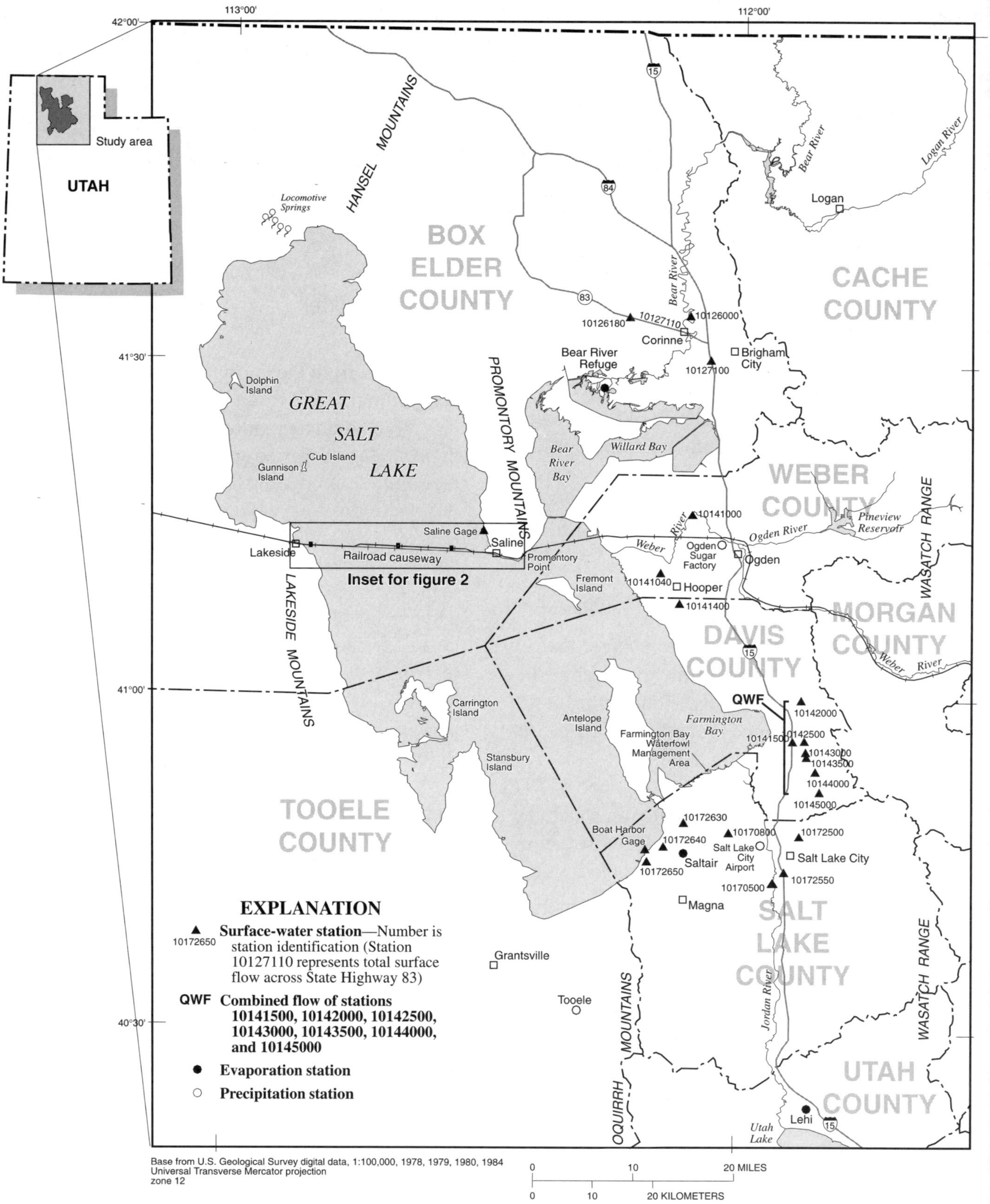

Figure 1. Location of study area and data-collection sites used to estimate inflow, water-surface altitude, and evaporation for Great Salt Lake, Utah, 1980–86. Modified from Waddell and Fields (1977, fig. 6).

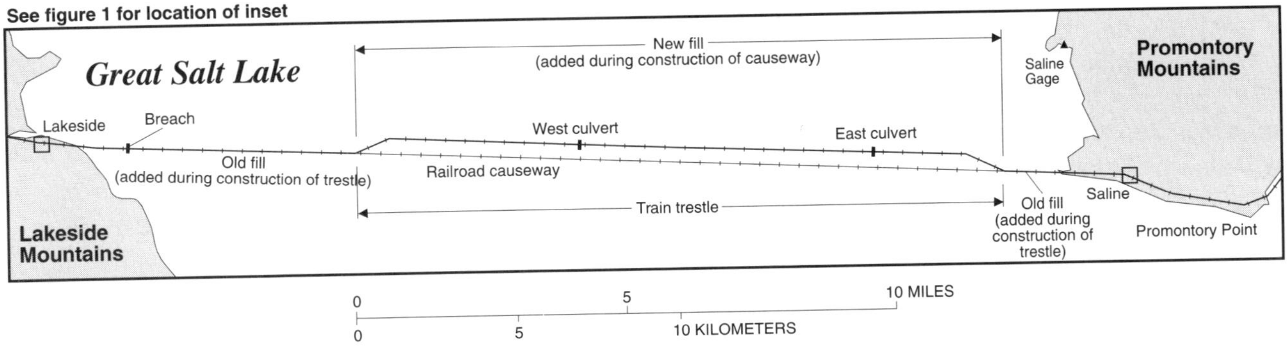

Figure 2. Location of breach and culverts in the causeway across Great Salt Lake, Utah.

of the causeway on the salt balance for variable culvert widths and tributary inflows to the lake. Because of the high water-surface altitude and modifications of the causeway conveyance properties during 1980–86, the original causeway model was no longer valid. The conveyance properties of the causeway originally included two culverts, each 15 ft wide, and the permeable rock-fill material. During 1980–86, fill material was frequently added to the causeway (fig. 2) to maintain the top of the causeway above the water surface; during 1983, the two 15-ft-wide culverts became submerged beneath the rising water surface and eventually filled with debris; and in 1984, a 300-ft-wide breach was constructed in the causeway. Because these new conditions warranted revision of the original causeway model, the U.S. Geological Survey (USGS), in cooperation with the Utah Department of Natural Resources, Division of State Lands and Forestry, completed this study to modify and improve the capability of the original causeway model for simulating the water and salt balance between the south and north parts of Great Salt Lake, Utah.

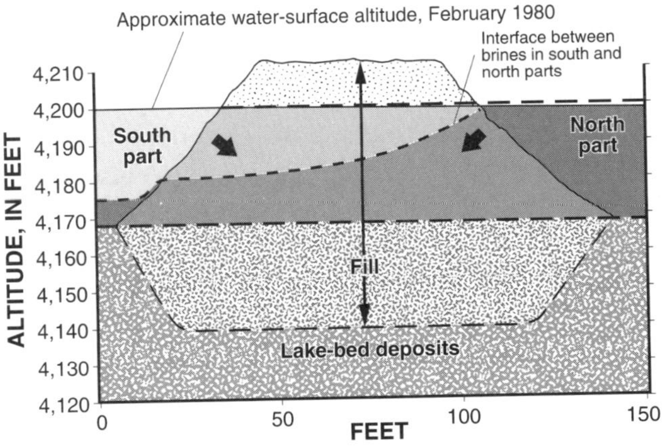

Figure 3. Diagrammatic cross section of the causeway across Great Salt Lake, Utah.

Description of the Study Area

Great Salt Lake is a closed lake located in semi-arid northwestern Utah in the Basin and Range Physiographic Province (Fenneman, 1931). The lake is bordered on the west by desert and on the east by the metropolitan Salt Lake City area at the base of the Wasatch Range. Great Salt Lake is a remnant of freshwater Lake Bonneville, which existed about 10 to 15 thousand years ago, covered much of western Utah and small parts of Idaho and Nevada, and was about 1,000 ft deep. In 1963, when Great Salt Lake was at its lowest water-surface altitude in recent history at 4,191.35 ft, it covered about 950 mi^2 and had a maximum depth of about 25 ft. In 1986, when Great Salt Lake was at its highest water-surface altitude in recent history at 4,211.85 ft, it covered about 2,400 mi^2 and had a maximum depth of about 45 ft.

Purpose and Scope

This report presents the results of the study that was done to modify the causeway model developed by Waddell and Bolke (1973). The main body of this report contains the results of the study, and the appendixes contain the details pertaining to the modification of the major components of the model. Included is a description of the calibration of the model and the results of simulations for three hypothetical 10-year periods in which the water-surface altitude decreased, remained at or near a constant level, or increased.

The model developed by Waddell and Bolke (1973) was modified so that it would be valid for water-surface altitudes up to 4,212 ft and for the changes in conveyance properties of the causeway during 1980–86. The major components of the original causeway model that were modified are the equations and subroutines used to compute the bidirectional, stratified flows through the causeway fill, culverts, and breach.

The equations used by Waddell and Bolke (1973) for computing flow through the culverts and breach were replaced with equations from Holley and Waddell (1976). Flows computed using the equations of Holley and Waddell were evaluated by comparing them with measured flows through both of the culverts and the breach.

The subroutines used to compute flow through the permeable fill were revised using the two-constituent solute-transport model (referred to in this report as the fill-flow model) of Sanford and Konikow (1985). The hydraulic properties of the fill material used as input to the fill-flow model were estimated from Waddell and Bolke (1973, p. 23), and no additional field investigation was done as part of this study. Seepage through the causeway fill was computed using the fill-flow model for a wide range of boundary conditions and was used as a dynamic component of the causeway model. The seepage computed by the fill-flow model was indirectly evaluated and revised by comparing the water and salt balance computed by the causeway model with independent computations of the water and salt balance from measured data.

Freshwater inflow and evaporation were compiled for Great Salt Lake for monthly intervals during 1980–86, and the data were used as input to the causeway model. The water and salt balances were computed from monitoring data collected during 1980–86. Dissolved salt load for the south and north parts of Great Salt Lake during 1980–86 was computed and used to evaluate the causeway model for the calibration period, 1980–86.

WATER AND SALT BALANCE OF GREAT SALT LAKE

Prior to construction of the causeway during 1957–59, the hydrologic characteristics of Great Salt Lake were typical of a closed lake. The water surface rose and fell in response to the balance between the amount of water evaporated from the surface and the amount of water contributed to the lake by runoff, ground-water inflow, and precipitation on the surface. During periods when the water surface fell, surface area and volume decreased and dissolved-solids concentration increased. During periods when the water surface rose, surface area and volume increased and dissolved-solids concentration decreased. The lake was thought to be well mixed and to have no stratification, and salt precipitation was thought to occur throughout the entire lake when the lake volume was small enough to exceed the saturation level for sodium chloride.

Effects of Causeway Construction

The railroad causeway was completed in 1959. Conveyance properties included two 15-ft-wide culverts and the hydraulic properties of the causeway rock-fill material. The effect of the causeway construction was to change the water and salt balance of Great Salt Lake by creating two separate but interconnected parts of the lake with different water-surface altitudes and densities. Changes in the water and salt balances from completion of the causeway in 1959 through 1986 are shown in figures 4 and 5.

Since completion of the causeway, the major independent factors that affect the water and salt balance of Great Salt Lake are surface-water inflow to the south part of the lake (*SIS*) and the conveyance properties of the causeway (fig. 6). Flow through the causeway, *QS* and *QN*, is primarily dependent on surface-water inflow and conveyance properties of the causeway. Because almost all surface-water inflow is to the south part, the water surface of the south part (*ES*) is at a higher altitude than that of the north part (*EN*) and the water in the south part is less saline than the water in the north part. As a result of the differences between the water-surface altitudes and densities of the south and north parts, brine flows in both directions through the causeway (*QS* and *QN*, figs. 6 and 7). Less dense brine from the south part flows northward through the upper part of the causeway and more dense brine from the north part flows southward through the lower part of the causeway.

The water and salt balances of the lake (figs. 6 and 7) respond dynamically to seasonal and long-term changes in surface-water inflow. Evaporation (*EOS* and *EON*) and precipitation (*PIS* and *PIN*) also affect the water and salt balances, but because the differences in the rates of evaporation and precipitation for the south and north parts are not large, these factors are of less importance.

Freshwater inflow to Great Salt Lake contributes a relatively insignificant part of the total salt load (dissolved and precipitated salt load) contained in the lake; thus, the total salt load can be assumed to be constant for a large number of years. Hahl (1968, p. 20) determined the salt load contributed from freshwater inflow

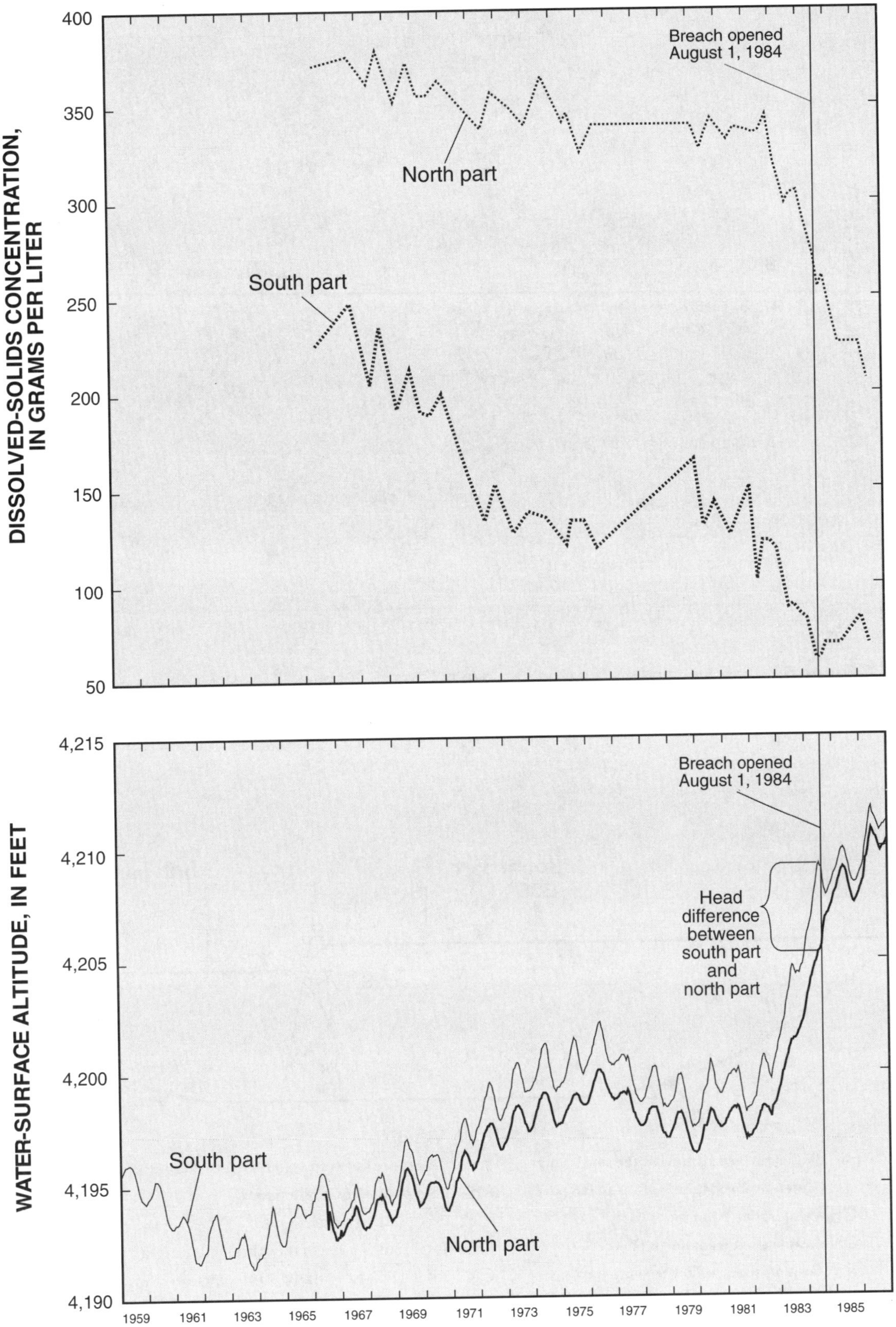

Figure 4. Dissolved-solids concentration, water-surface altitude, and head difference of brine in the south and north parts of Great Salt Lake, Utah, 1959–86.

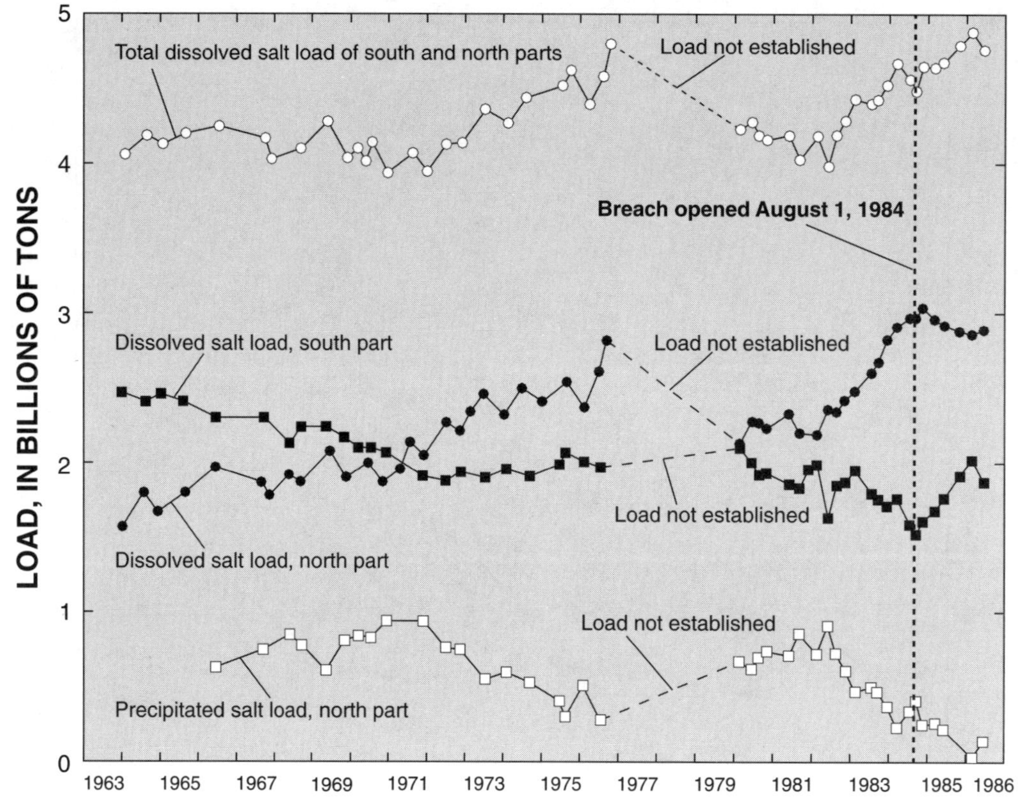

Figure 5. Dissolved and precipitated salt load in Great Salt Lake, Utah, 1963–86.

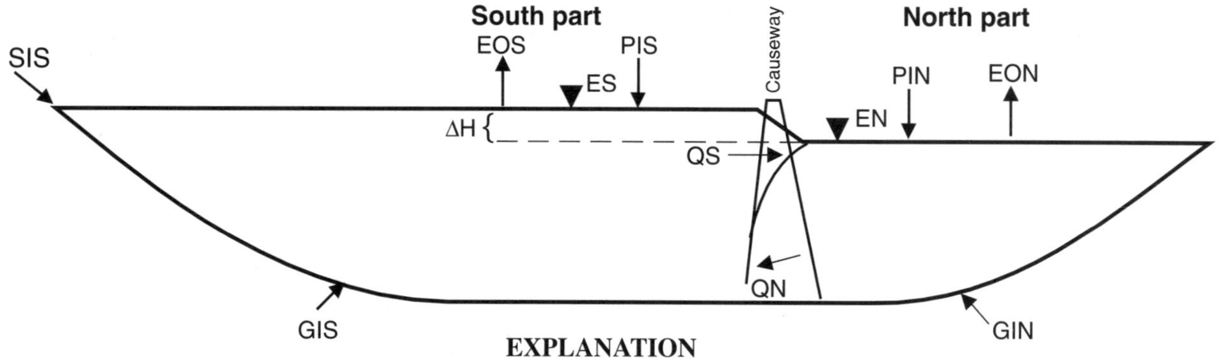

EXPLANATION

ES	Water-surface altitude of the south part	ΔH	Difference between water-surface altitude of south and north parts
EN	Water-surface altitude of the north part	PIS	Precipitation on south part
EOS	Evaporation from south part	PIN	Precipitation on north part
EON	Evaporation from north part	QS	Total south-to-north flow through causeway
GIS	Ground-water inflow to south part	QN	Total north-to-south flow through causeway
GIN	Ground-water inflow to north part	SIS	Surface-water inflow to south part

Figure 6. Schematic diagram of water balance for Great Salt Lake, Utah.

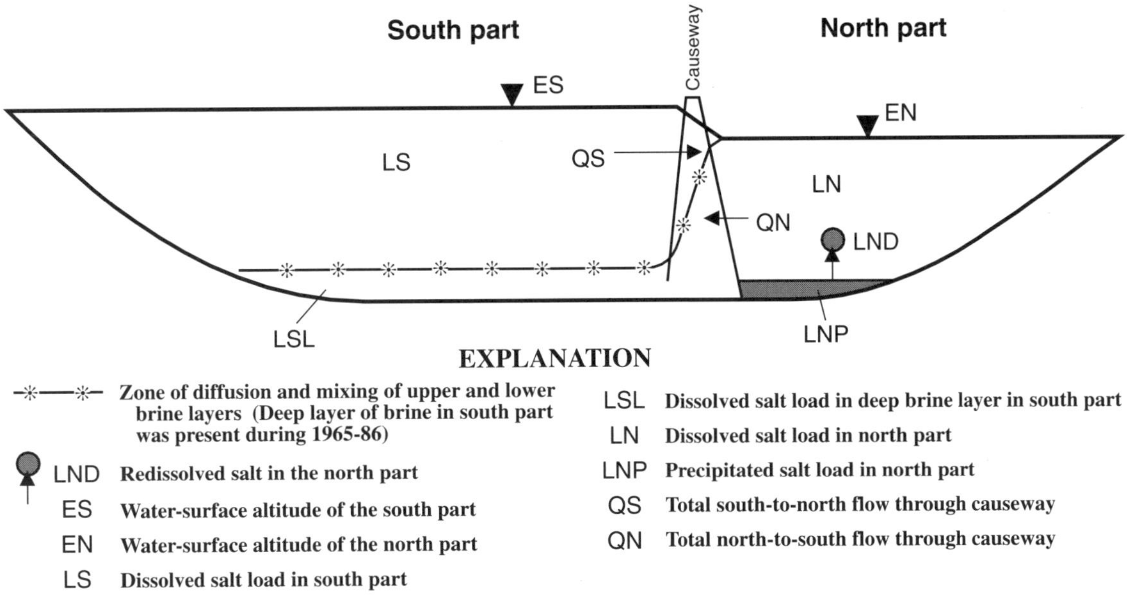

Figure 7. Schematic diagram of salt balance for Great Salt Lake, Utah.

to be less than 0.0035 billion tons per year. During 1980–86, the total salt load was determined to be about 4.9 billion tons (see Appendix B). Therefore, during a 100-year period, the total salt load of the lake would increase by about 0.35 billion tons, or about 7 percent.

The trend of dissolved salt load for the south and north parts (*LS* and *LN*) during 1963–86 provides evidence of the effects of the causeway on the balance of dissolved salt load between the south and north parts (fig. 5). The trend of total dissolved salt load (fig. 5) can be used as an indication of salt precipitation (*LNP*) or re-solution (*LND*). Because the total amount of salt available in the lake is assumed to be constant for a large number of years, an increase in total dissolved salt load represents re-solution of salt, and a decrease in total dissolved salt load represents precipitation of salt.

In 1963, shortly after completion of the causeway, the water surface declined to its lowest recorded altitude (4,191.35 ft) and volume. At this low volume, the south and north parts of the lake were saturated with respect to sodium chloride, and a salt crust formed on the lake bed south and north of the causeway (Madison, 1970, p. 12).

During 1964–71, the water surface of the lake generally rose (fig. 4) and the south part freshened because of increasing lake volume and the net movement of the dissolved salt load to the north part (fig. 5). The load loss from the south part ceased about 1972, indicating that the balance of dissolved salt loads between the south and north parts was near equilibrium for the inflow conditions and causeway conveyance properties existing during 1964–72. This equilibrium is indicated by the comparatively constant dissolved salt load in the south part (fig. 5) during 1972–80. The dissolved salt load in the north part increased during 1972–76 because of re-solution of the salt crust in the north part as the water surface rose and the volume of the lake increased. Most of the salt that precipitated throughout the entire lake during 1959–63 probably dissolved from the south part by 1972, but the precipitated salt in the north part could not redissolve because the brine in the north part was at or near saturation.

A lower layer of brine with relatively constant volume was first observed in 1965 in the south part of the lake (*LSL*). This layer had chemical characteristics similar to the brine in the north part. Madison (1970, p. 12) observed that the lower layer of brine in the south part of the lake occurred everywhere the lake bottom is below 4,175 ft. Stratification in the south part occurred because the causeway reduced circulation in the lake. The causeway created the conditions necessary for the north part to achieve and maintain a higher density than the south part. Because of the higher density in the north part, brine could flow from the north to south part through the culverts and permeable fill, thereby providing a constant supply of high-density brine and maintaining the deep south part brine layer. Data collected by the USGS and the Utah Geological Survey (UGS) during 1973–74 indicated that the volume of the lower layer of brine in the south part and the altitude of the interface with the overlying brine was essentially unchanged, even though the water surface had increased in altitude by several feet (Appendix B, section entitled "Stratification in Great Salt Lake").

During 1980–86, annual inflow averaged 4,627,000 acre-ft, and the water surface rose 14.2 ft, reaching a historic high altitude of 4,211.85 ft on June 3, 1986. Average inflows for 1980–86 were about 240 percent greater than the long-term average of 1,926,000 acre-ft (Waddell and Barton, 1980, p. 11) that was estimated for 1931–76.

Modifications to the conveyance properties of the causeway and record inflows to the lake during 1980–86 affected the salt balance. The causeway fill was raised and widened during 1980–86 because of the rising water surface. The addition of the fill material probably changed the hydraulic characteristics of the fill.

During January 1, 1980, to July 31, 1984, total south-to-north flow through the causeway averaged about 4,900 ft^3/s and total north-to-south flow averaged about 1,600 ft^3/s (table 1). The record inflow caused about 0.5 billion tons of dissolved salt to move from the south to the north part. The overall net decrease in dissolved-solids concentration in the south part of the lake was 110 g/L, 18 g/L as a result of net load movement from south to north and 92 g/L as a result of an increase in volume. During this time, the water-surface altitude of the south part rose 11.4 ft (figs. 4 and 5).

The dissolved salt load of the north part also increased as a result of precipitated salt dissolving in the north part. The precipitated salt dissolved because the increase in the volume of the lake caused the dissolved-solids concentration in the north part to decrease below saturation. From 1972 through July 1984, the volume of the deep layer of brine in the south part and the altitude of the interface with the overlying brine increased, and the interface in the south part became more diffused. There was little or no stratification in the north part prior to 1984, but the north part became stratified by July 1984 as a result of the rapidly increasing depths caused by record inflows and water-surface altitudes (Appendix B, section entitled "Stratification in Great Salt Lake"). After about 1991, both the south and north parts became well mixed with no stratification (J.W. Gwynn, Utah Geological Survey, written commun., 1994).

Effects of Causeway Breach

To combat the rising level of Great Salt Lake in 1984, the State of Utah completed construction of a 300-ft opening (breach) on the western edge of the lake (fig. 2). The breach was designed to equalize the head differential that had developed between the north and south parts. After the 300-ft-wide breach was opened on August 1, 1984, total south-to-north flow during August 1, 1984, to December 31, 1986, averaged about 7,400 ft^3/s and total north-to-south flow averaged about 3,500 ft^3/s (table 1). Compared with average

Table 1. Flow through fill, culverts, and breach of causeway across Great Salt Lake, Utah, 1980–86

[Flow in cubic feet per second. Fill: Flow computed using causeway model; Culvert: Annual flow compiled from flow measurement made by U.S. Geological Survey (1980, 1981) and ReMillard and others (1982, 1983); No culvert flow is inferred from April 1, 1983, through December 31, 1986, as a result of the high water-surface altitude causing difficulty in cleaning debris from culverts; Breach: Flow began August 1, 1984, and annual flow was compiled from measurements by ReMillard and others (1984, 1985, 1986)]

Type of flow	January 1, 1980, through July 31, 1984		August 1, 1984, through December 31, 1986	
	South to north	North to south	South to north	North to south
Fill	3,300	1,400	2,200	3,000
Culvert	1,600	200	0	0
Breach	0	0	5,200	500
Total	**4,900**	**1,600**	**7,400**	**3,500**

flow during January 1980 to July 1984, the effect of the breach was to increase south-to-north flow by about 50 percent and more than double north-to-south flow. Because of the increased flow through the causeway, the head difference between the south and north parts decreased from a range of about 3.0 to 3.9 ft prior to the breach to 0.5 to 1.0 ft after the breach.

The increased flow in both directions through the causeway caused the dissolved salt load generally to increase in the south and decrease in the north. From August 1, 1984, when the breach was opened, to December 31, 1986, a net dissolved salt load of about 0.3 billion tons moved from the north to south part (fig. 5). The overall increase in dissolved-solids concentration in the south part was 8 g/L. The net increase in concentration resulted from an increase of 19.5 g/L from a net load movement from north to south and a decrease of 11.5 g/L from a net increase in volume in the south part. During this time, the water-surface altitude of the south part rose 2.8 ft (fig. 3).

The concentration of dissolved solids in the north part decreased below saturation (355 g/L), and most precipitated salt in the north part is believed to have dissolved. The density stratification for the north and south parts of the lake changed after the breach was opened. The increased circulation caused the sharp interfaces between the upper and lower layers of brine in the south and north parts to become less diffused (Appendix B, section titled "Stratification in Great Salt Lake").

SIMULATION OF WATER AND SALT MOVEMENT THROUGH THE CAUSEWAY

The original causeway model developed by Waddell and Bolke (1973) was used extensively during the 1970s by State agencies to simulate water and salt movement through the causeway and to determine the effects of causeway modifications on the salt balance between the two parts of the lake. The original causeway model also was used to simulate the effects of different widths of culvert or breach openings on the head difference between the south and north parts of the lake. The simulations made by the original causeway model were used to help design the breach that was constructed in the causeway in 1984 to help alleviate flooding along the shores of the south part of the lake.

The causeway model simulates the effects of the causeway on the water and salt balance of the lake for variable rates of inflow. The main components of the causeway model and modifications to the original model are shown in figure 8. The causeway model has all the simulation capabilities of the original model, but has the additional capability of simulating the effects of the breach.

Calibration of Causeway Model

Calibration of the causeway model was done by comparing the head differences and dissolved and precipitated salt loads simulated by the causeway model with those computed from data measured during 1980–86 (fig. 9). The difference between the simulated and measured values represents the combined error from the causeway-fill, culvert, and breach flow.

For the 1980–86 calibration period, the simulated head differences are smaller than the measured head differences, the simulated precipitated salt load in the north part is smaller than the measured precipitated salt load, and the simulated dissolved salt load of the south part is larger than the measured dissolved salt load. Each of these differences is consistent with the occurrence of too much circulation through the causeway, which could be caused by the computed flow in both directions being too large.

During the calibration period, 1980–86, there were periods when either the causeway-fill, culvert, or breach flow was the principal flow through the causeway. Error between the simulated and measured values was noted to be large during some parts of the calibration period. By associating the amount of error with the time when a specific mode of conveyance was contributing most of the flow, the most likely cause of the error was attributed to error in the computed values of fill flow.

In 1980, when flow through the causeway was conveyed only through the culverts and fill, the simulated values of head difference, dissolved salt loads in the south part, and precipitated salt loads in the north part began deviating from the measured values. These errors between the simulated and measured values generally increased until the breach was opened on August 1, 1984 (fig. 9).

In April 1983, the culverts became submerged, and removal of the rock debris that frequently fills the culverts during periods of high winds ceased. Prior to 1993, the culverts were last cleaned in 1983 (Orlando Miera, Southern Pacific Transportation Co., oral commun., 1993). The culverts probably became either partly or completely blocked between about February 1984 and July 1984; therefore, most brine was proba-

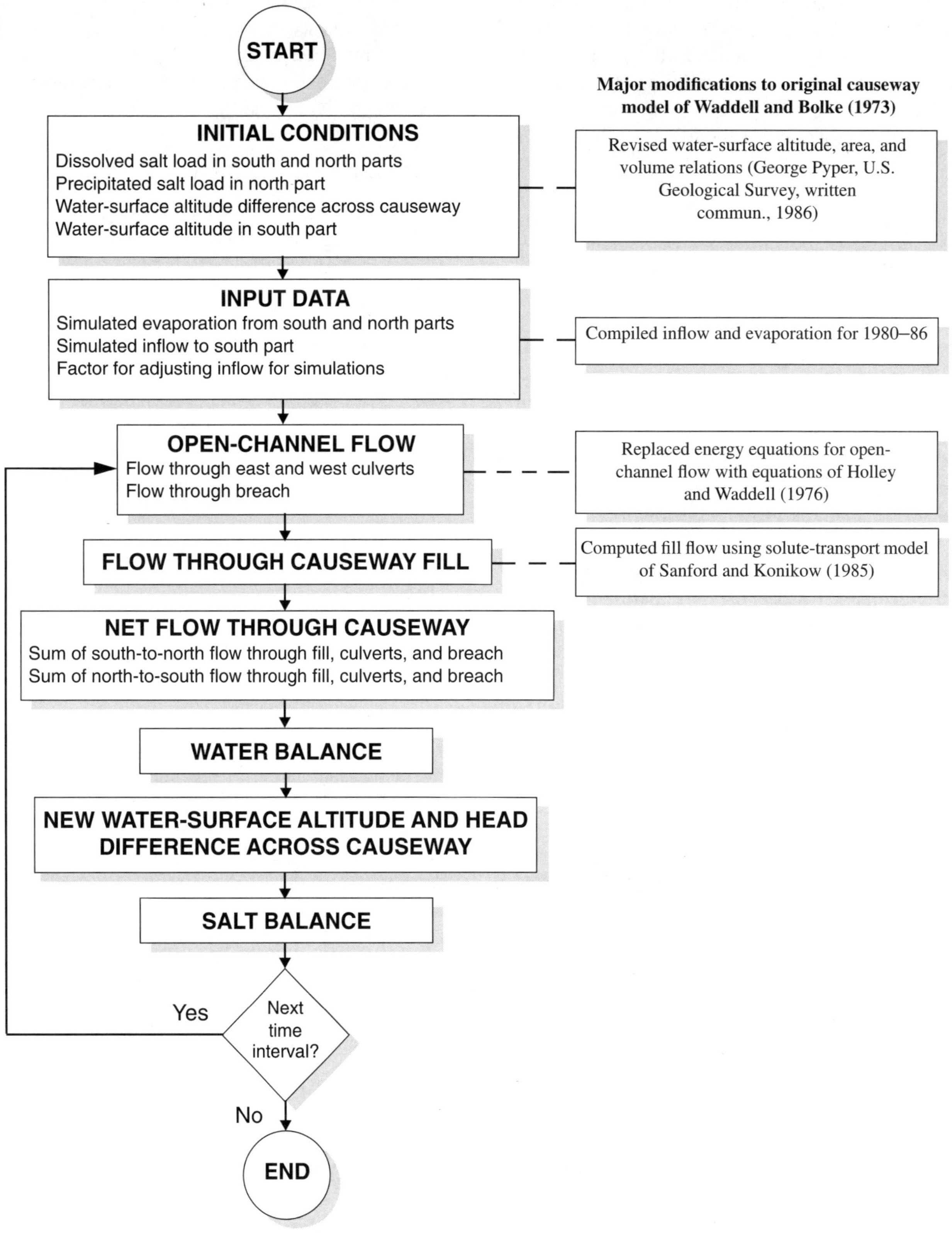

Figure 8. Flow chart of causeway model for Great Salt Lake, Utah, showing major modifications made to original causeway model of Waddell and Bolke (1973).

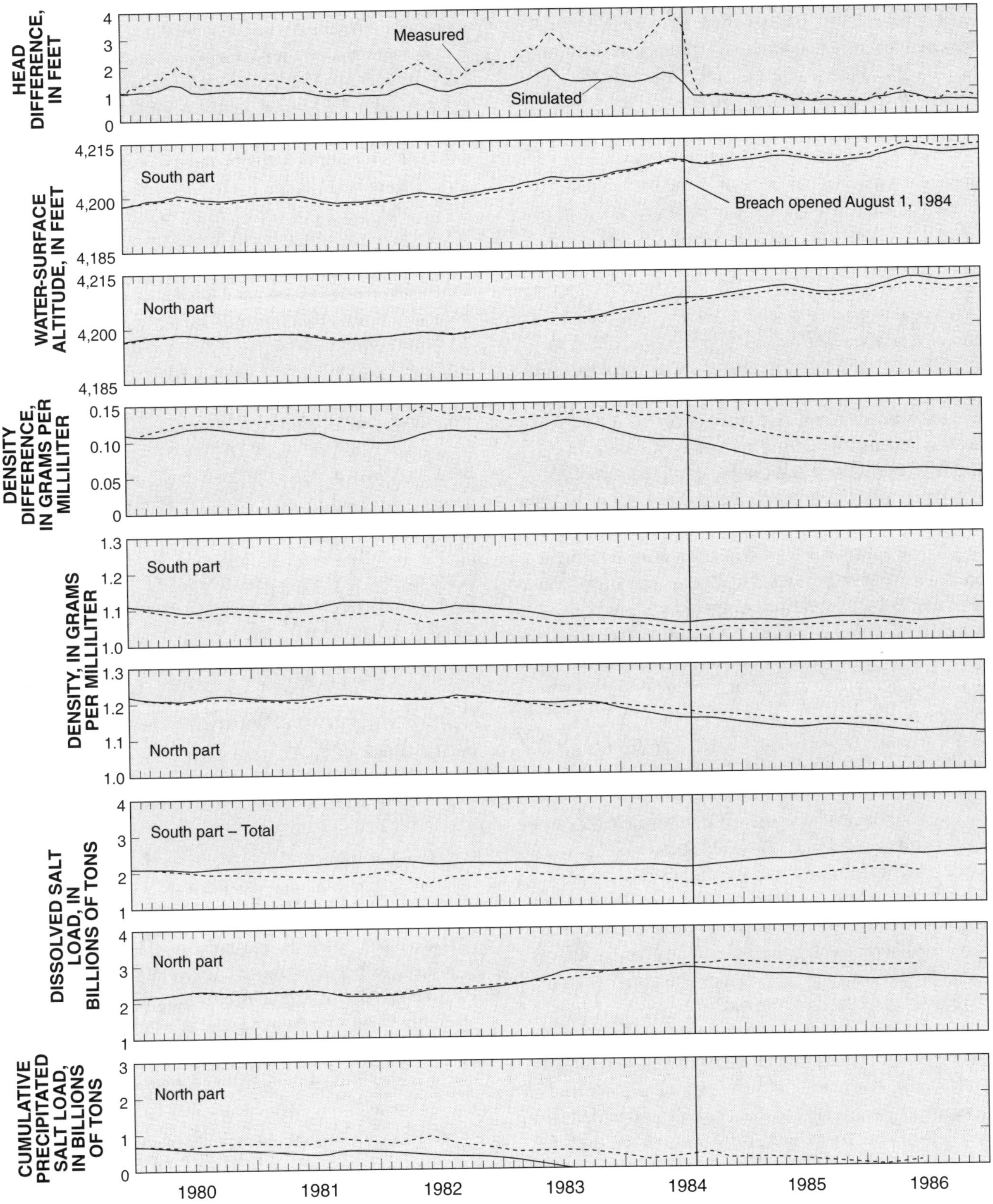

Figure 9. Simulated and measured head difference, water-surface altitude, density difference, density, and dissolved salt and cumulative precipitated salt load in the south and north parts of Great Salt Lake, Utah, 1980–86, before calibration of flow through the causeway fill.

bly conveyed only through the fill during this period. During this period, there was an increase in error between the simulated and measured head difference (fig. 9). Because simulated head difference was smaller than measured head difference, the simulated fill flow was probably too large.

From August 1, 1984, when the breach was opened, to December 1986, only the breach and fill conveyed flow. Because the breach was about 300 ft wide, total flow from south to north was dominated by flow through the breach. The increase of flow resulting from the breach (table 1) caused measured head differences to decrease from about 3.9 ft to less than 1.0 ft and caused the error between the simulated and measured values of head difference to become small relative to the error noted prior to the breach opening. Because the small relative error occurred after the breach became the dominant means of conveying south-to-north flow, error in the computation of fill flow was most likely the cause of the larger error prior to the breach.

The subroutines used to calculate culvert and breach flow were evaluated independently of the causeway model by comparing computed with measured flow for observed boundary conditions (Appendixes D and E). Fill flow could not be directly evaluated with the same methods as culvert or breach flow, so different aspects of the data used to estimate fill flow were compared with the values calculated by the fill-flow model (Appendix C, section entitled "Methods used to compare causeway-fill flow").

Field data used in the fill-flow model were collected during 1971–72. The field studies and data collection included time-of-travel studies for two sets of boundary conditions and borehole data collected from drilling 10 test holes (Waddell and Bolke, 1973, p. 20). A comparison of velocities determined from time-of-travel studies and fill-flow-model-computed velocity profiles is discussed in Appendix C.

The causeway fill was raised and widened during the 1980s, but no additional field studies were done to evaluate the hydraulic properties of the fill material. Because of the changes to the causeway fill and the limited sets of data to estimate flow through the original fill, there may be considerable error in the estimates of causeway-fill flow.

When the field studies were done on the causeway fill during 1971–72, there was considerable range in the hydraulic-conductivity values at the 10 sections along the 12.21-mi reach of the central part of the causeway where time-of-travel studies were done. To simplify modeling of the causeway-fill flow (Appendix C), average values of hydraulic conductivity and porosity were used for a single cross section that was considered to be representative of the entire causeway.

Assuming that inaccuracies of the hydraulic-conductivity and/or porosity values used in the fill-flow model were responsible for the errors between the simulated and measured values of head difference and dissolved and precipitated salt loads, flow computed by the fill-flow model was reduced until the minimum error between simulated and measured values was attained. Reducing fill flow to 40 percent for the 1980–86 calibration period created the best match between simulated and measured values (fig. 10). No changes were made to any of the other components of the causeway model as part of the calibration.

After calibration of fill flow, maximum deviation between the dissolved salt load as simulated by the causeway model and that computed from water-quality data was 0.34 billion tons for the south part and 0.17 billion tons for the north part. The standard error of estimate as a percentage of the mean for dissolved salt loads simulated by the causeway model and those measured was 5 percent for the south part and 4 percent for the north part.

Constraints and Assumptions for the Simulated Period

The constraints and assumptions for the simulated period are as follows:

- The optimum operating range of the causeway model for water-surface altitudes is from 4,191 to 4,212 ft.
- The causeway model is limited to positive head differences between south and north parts and positive density differences between north and south parts. The causeway model is valid for head differences ranging from about 0.1 to 3.9 ft and density differences ranging from about 0.02 to 0.15 g/mL.
- The causeway model is capable of simulating water and salt movement across the causeway using any combination of up to two culverts or breaches combined in the model to make one wide culvert.
- The simulations were made with the assumptions that the altitude of the culvert and breach bottoms remained constant.

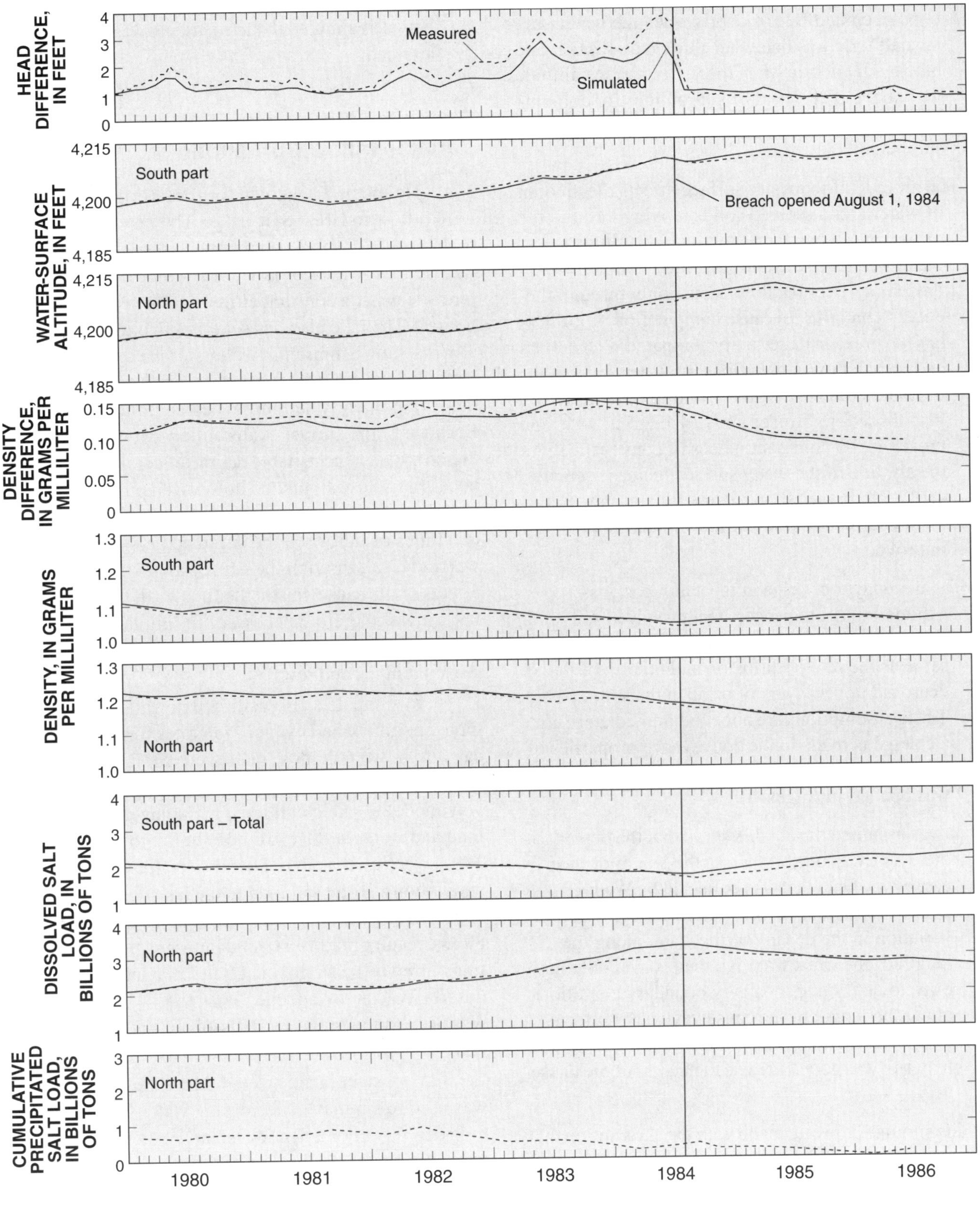

Figure 10. Simulated and measured head difference, water-surface altitude, density difference, density, and dissolved salt and cumulative precipitated salt load in the south and north parts of Great Salt Lake, Utah, 1980–86, after calibration of flow through the causeway fill.

- The crown of the east culvert was estimated to be at an altitude of 4,203 ft. The culverts were assumed to be open and free of debris when the water-surface altitude was below an altitude of 4,203 ft and blocked by debris when the water-surface altitude was above 4,203 ft. The subroutines for computing culvert flow are not valid for flow through a submerged culvert.
- The altitude of the water surface affects the amount of water and dissolved solids moving through the causeway. When the water-surface altitude is less than or equal to the altitude of the bottom of the breach (4,199.5 ft), flow occurs only through the culverts and fill. Because computation of fill flow has the greatest uncertainty compared with culvert and breach flow, simulations for water-surface altitudes less than 4,199.5 ft would have more error than those for altitudes above 4,199.5 ft, when the breach is the dominant means of conveyance. As the altitude of the water surface increases above 4,199.5 ft, breach flow becomes increasingly dominant and the accuracy of the simulation is improved.
- The deep layer of brine in the south part was assumed to contain a constant dissolved salt load, and the rate of north-to-south flow is about the same as the rate of diffusion and mixing of the deep and upper layers of brine in the south part. These assumptions are not valid immediately after a change is made in the conveyance properties of the causeway, such as occurred when the breach was opened in August 1984.
- It was assumed during this study that the density of the water was uniform along the length of the causeway. Data that was collected by Waddell and Bolke (1973, table 8) showed that there was little variation in the density of the water along the length of the causeway. No data is available however, to determine if density boundary conditions along the length of the causeway were uniform after construction of the breach or after more fill material was added to raise the height of the causeway.
- The simulations were made with the assumption that inaccuracies of the hydraulic-conductivity and/or porosity values used in the fill-flow model were responsible for the errors between the simulated and measured values of head difference and dissolved and precipitated salt loads, so that flow computed by the fill-flow model was reduced by a series of constant factors until the minimum error between simulated and measured values was attained.

Simulated Movement of Water and Salt for Variable Inflow Conditions

No method is known for predicting future long-term inflow to Great Salt Lake. The lake-level hydrograph for 1850–1986 indicates that the water surface generally rises or falls for a period of years and rarely stays at a constant altitude for an extended period. The lake also has seasonal fluctuations, generally reaching a minimum water-surface altitude during the fall when net inflow is low and a maximum during the spring or early summer when net inflow is high. Because future surface-water inflow cannot be predicted with any acceptable degree of confidence, it was necessary to use simulated inflow as input to the causeway model and then to evaluate the effects of different quantities of inflow on the simulated movement of water and salt through the causeway. To simulate the effects of the causeway on the future salt balance in the lake, it was therefore necessary to simulate both long-term rising or falling water-surface trends as well as seasonal highs and lows.

To simulate a 10-year trend in water-surface altitude, net inflow and evaporation rates for 1981 (a typical year in which inflow rates were similar to evaporation rates) were used as input data to the causeway model. Lake conditions, including dissolved salt load and water-surface altitude similar to those during 1991, were used as initial conditions for the 10-year simulations. Total salt load in the lake decreased from about 4.9 billion tons to 4.3 billion tons during 1987–89 as a result of the West Pond pumping project, which transferred brine from Great Salt Lake into the west desert (Wold and Waddell, 1994); thus, 4.3 billion tons was used as the initial total dissolved salt load in the lake.

A constant factor (*IR*) of 0.55 times net inflow was used to simulate a decreasing water-surface altitude (fig. 11). For a near constant water-surface altitude, *IR* equals 0.9 (fig. 12), and for an increasing water-surface altitude, *IR* equals 1.5 (fig. 13). Because the inflow rate is multiplied by a constant factor from the beginning to the end of the simulated period, most of the increase or decrease in water-surface altitude occurs during the first few years of the simulated period.

For simulated inflow rates during a hypothetical 10-year period with the culverts and breach open that resulted in the water-surface altitude decreasing from about 4,200 to 4,192 ft (fig. 11), there was a net movement of about 1.0 billion tons of dissolved salt from the south to the north part, and about 1.7 billion tons of salt precipitated in the north part. For simulated inflow rates during a hypothetical 10-year period with the culverts and breach open that resulted in a near-constant water-surface altitude of about 4,200 ft (fig. 12), there was a net movement of about 0.4 billion tons of dissolved salt from the south to the north part, and about 0.2 billion tons of salt precipitated in the north part of the lake. For simulated inflow rates during a hypothetical 10-year period with the culverts and breach open that resulted in the water-surface altitude increasing from about 4,200 to 4,212 ft (fig. 13), there was a net movement of about 0.2 billion tons of dissolved solids from the south to the north part, and no salt precipitated in the north part of the lake.

The probability of such simulated water-surface trends actually occurring is small, but the range of simulated water surfaces probably incorporates the range of water surfaces that have occurred during 1963–86. In the past, long periods of increasing water-surface altitude included individual years in which the water-surface altitude decreased, and similarly, long periods of decreasing water-surface altitude included individual years when the water-surface altitude increased.

Salinity differences between the south and north parts during the simulated period may differ from those of 1980; consequently, evaporation rate also differed. Equations were used to compensate for the effects of changes of salinity on evaporation rate (Appendix A).

SUMMARY

The current water and salt balance of Great Salt Lake depends primarily on the amount of inflow from tributary streams and the conveyance properties of a causeway that divides the lake into the south and north parts. The causeway was constructed during 1957–59 and partially restricts circulation between the south and north parts. The conveyance properties of the causeway originally included two 15-ft-wide culverts and the permeable rock-fill material, but the causeway has since been modified. Waddell and Bolke (1973) described the effects of the causeway on the water and salt balance and developed a model for simulating the effects of the causeway on the salt balance for variable culvert widths and inflows.

During 1980–86, the salt balance changed as a result of record high inflows and modifications to the causeway conveyance properties. Annual inflow during 1980–86 averaged 4,627,000 acre-ft and the lake surface reached a historic high altitude of 4,211.85 ft. Average inflow for 1980–86 was about 240 percent greater than the long-term average of 1,926,000 acre-ft that was estimated for 1931–76.

Because of the high water-surface altitudes during 1980–86, fill material was added to the causeway. During 1983, the culverts became submerged and eventually filled with debris, and in 1984, a 300-ft-wide breach was constructed in the causeway. This study was done by the U.S. Geological Survey in cooperation with the Utah Department of Natural Resources, Division of State Lands and Forestry, to revise the model of Waddell and Bolke (1973) with the higher water-surface altitudes and modifications to the causeway. In this study, the water and salt balances were recomputed using monitoring data collected during 1980–86.

During January 1, 1980, to July 31, 1984, and prior to opening of the breach in the causeway, a net load of 0.5 billion tons of dissolved salt moved from the south to north part of the lake, primarily as a result of the record inflows. The breach was opened on August 1, 1984. During August 1, 1984, to December 31, 1986, a net load of 0.3 billion tons of dissolved salt moved from the north to south part of the lake.

Water-surface altitude affects the amount of water and dissolved solids moving through the causeway. The breach conveys water only when the water-surface altitude exceeds the altitude of the breach bottom, which is about 4,199.5 ft.

For simulated inflow rates during a hypothetical 10-year period with the culverts and breach open that resulted in the water-surface altitude decreasing from about 4,200 to 4,192 ft, there was a net movement of about 1.0 billion tons of dissolved salt from the south to the north part, and about 1.7 billion tons precipitated in the north part. For simulated inflow rates during a hypothetical 10-year period with the culverts and breach open that resulted in the water-surface altitude increasing from about 4,200 to 4,212 ft, net movement of dissolved salt to the south part was about 0.2 billion tons.

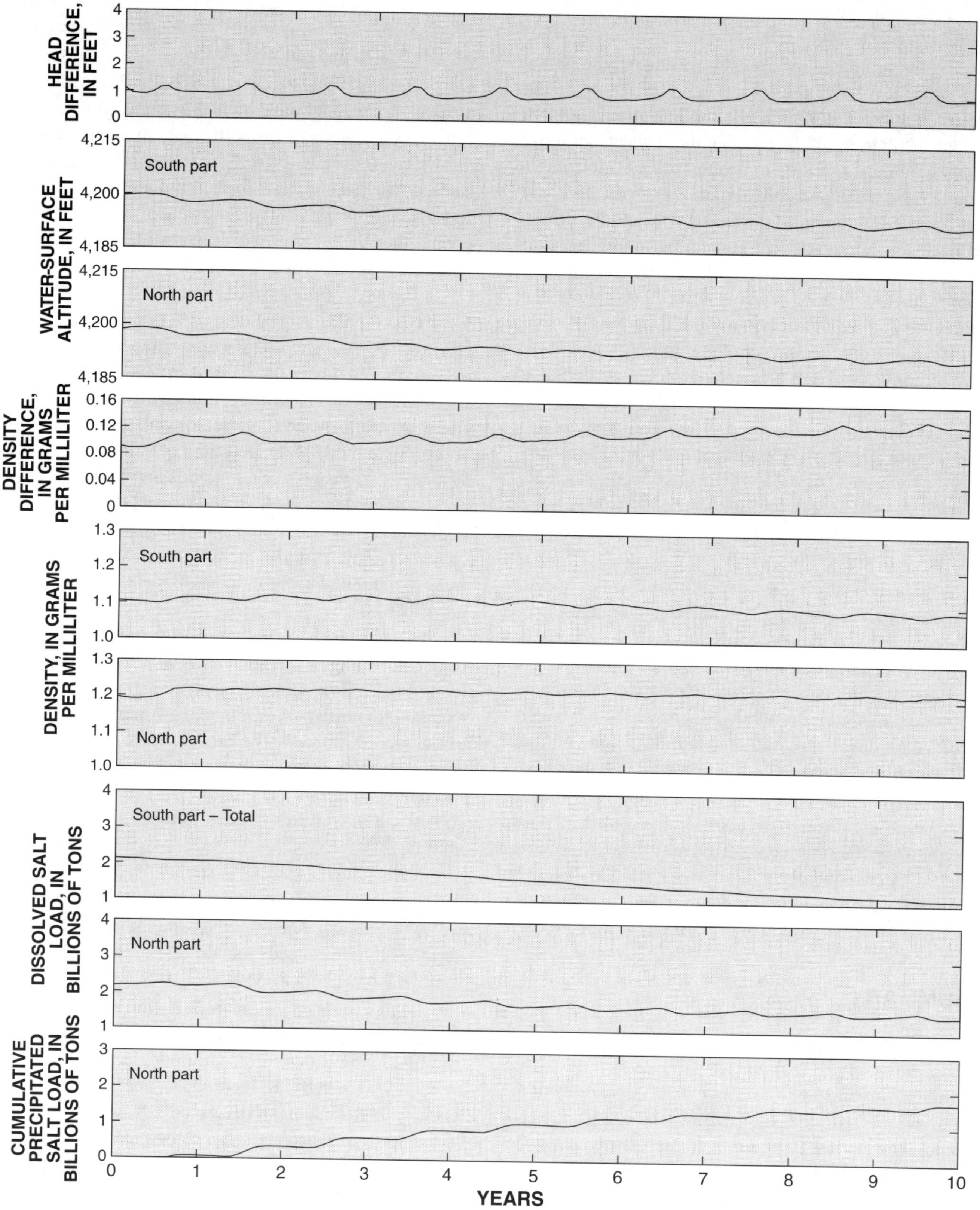

Figure 11. Effect of simulated inflow on the head difference, water-surface altitude, density difference, density, and dissolved salt and cumulative precipitated salt load in the south and north parts of Great Salt Lake, Utah, for decreasing water-surface altitude.

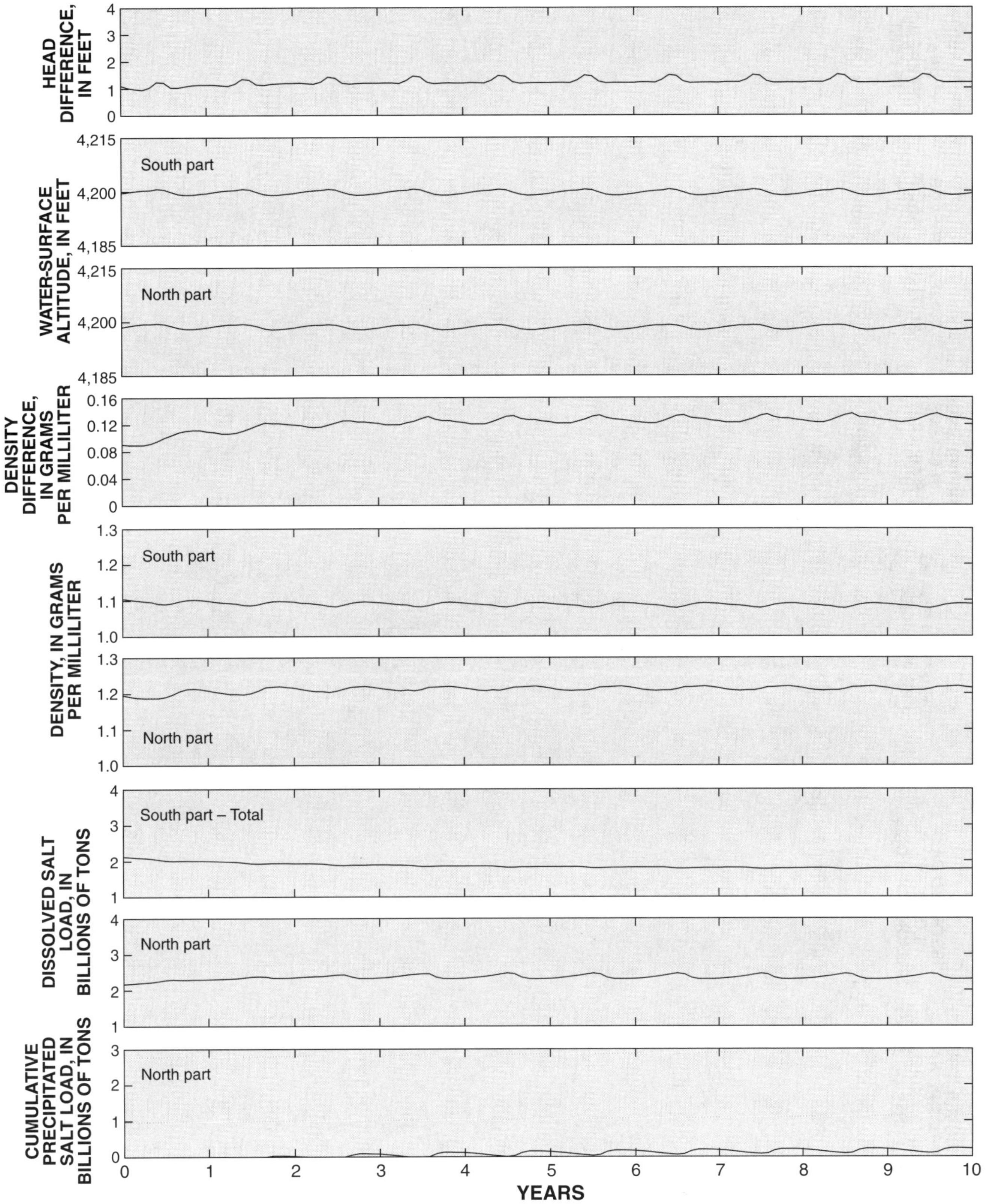

Figure 12. Effect of simulated inflow on the head difference, water-surface altitude, density difference, density, and dissolved salt and cumulative precipitated salt load in the south and north parts of Great Salt Lake, Utah, for constant water-surface altitude.

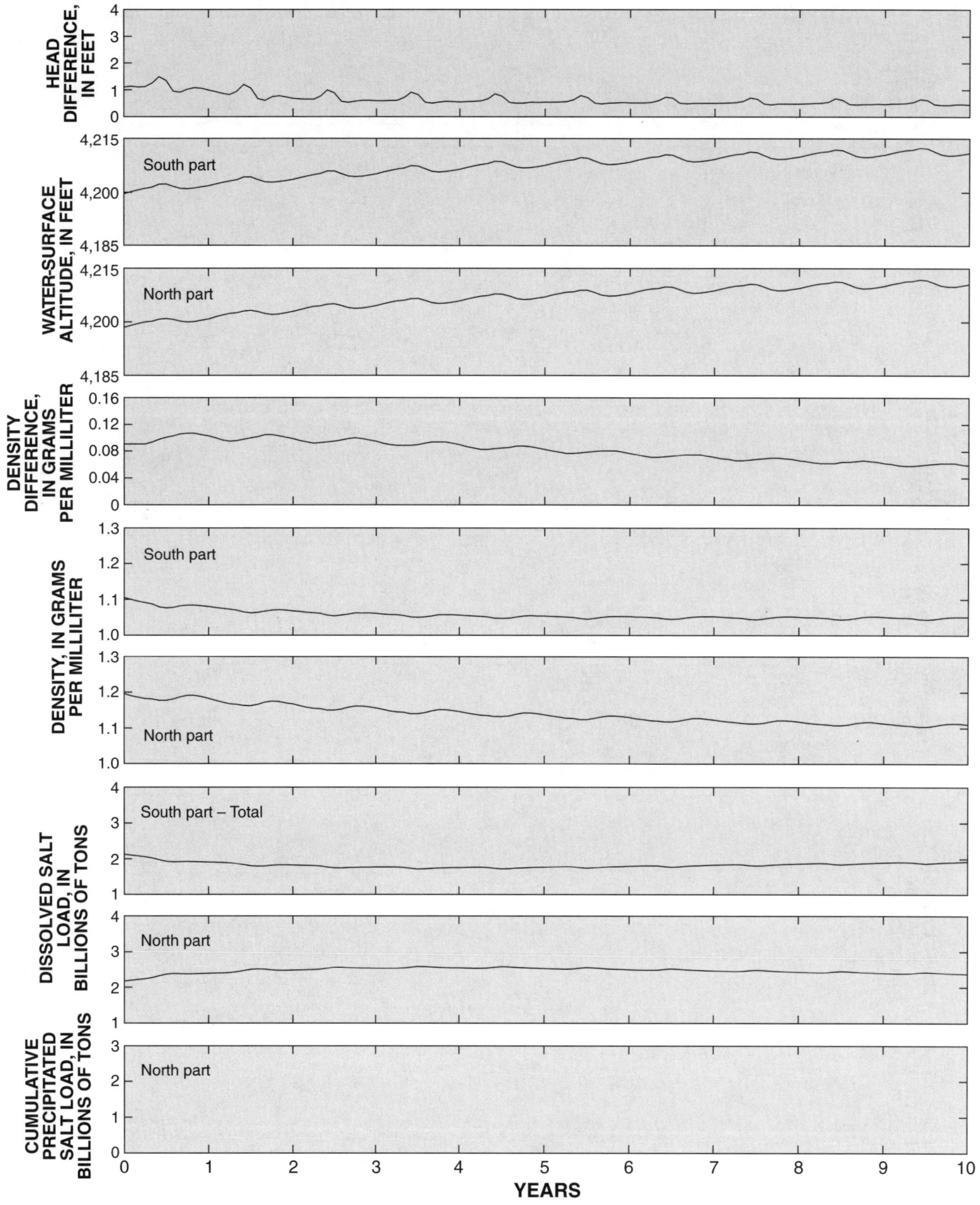

Figure 13. Effect of simulated inflow on the head difference, water-surface altitude, density difference, density, and dissolved salt and cumulative precipitated salt load in the south and north parts of Great Salt Lake, Utah, for increasing water-surface altitude.

INFORMATION NEEDED FOR GREATER ACCURACY

The causeway model developed during this study was based on data collected during 1980–86. The causeway model provides simulations for combinations of culvert and breach openings that are valid within the stated limits of accuracy. These simulations will be useful in planning future activities as well as to those concerned with modification of the causeway, which affects the net movement of dissolved salt. Greater accuracy for future simulations could be obtained by monitoring water-surface altitude and head difference across the causeway; measuring flow, specific gravity, and water-surface altitude monthly in the east and west culverts and breach; measuring bottom altitude of culverts and breach monthly; drilling test holes to determine hydraulic properties of the causeway fill; conducting tracer studies in the causeway fill; sampling both parts of the lake semiannually; determining the density variation of water along the length of causeway so that the boundary conditions can be accurately defined; and coring the salt crust at a few selected sites to detect changes in thickness. These data could be used to refine the causeway model.

Adding equations to the model for estimating culvert flow during submerged conditions and developing a circulation model capable of simulating currents in the lake also would enhance the accuracy of simulating water and salt movement through the causeway.

REFERENCES CITED

Adams, T.C., 1934, Evaporation from the Great Salt Lake: American Meteorological Society Bulletin, v. 15, p. 35–39.

Badon-Ghyben,W., 1888, Nota in Verband met de Voorgenomen Putboring Nabij Amsterdam (Notes on the probable results of well drilling near Amsterdam), Tijdsdar, Kon. Inst., Ing., The Hague, 1888/9, p. 8–22.

Fenneman, N.M., 1931, Physiography of the Western United States: New York, McGraw-Hill, 534 p.

Gwynn, J.W., 1988, Great Salt Lake brine sampling program 1985–1987: Utah Geological Survey Open-File Report 117, 19 p.

Gwynn, J.W., and Sturm, P.A., 1987, Effects of breaching the Southern Pacific railroad causeway, Great Salt Lake, Utah—Physical and chemical changes, August 1, 1984–July 1986: Utah Geological Survey Water-Resources Bulletin 25, 25 p.

Hahl, D.C., 1968, Dissolved-mineral inflow to Great Salt Lake and chemical characteristics of the Salt Lake brine: Summary for water years 1960, 1961, and 1964: Utah Geological Survey Water-Resources Bulletin 10, 35 p.

Herzberg, A., 1901, Die Wasserversorgung einiger Nordseebaden (The water supply on parts of the North Sea coast in Germany), Z. Gasbeleucht, Wasserversorg, 44, p. 815–819; 824–844.

Holley, E.R., and Waddell, K.M., 1976, Stratified flow in Great Salt Lake culvert: Journal of the Hydraulics Division, American Society of Civil Engineers, v. 102, no. HY7, Proceedings Paper 12250, July 1976, p. 969–985.

Madison, R.J., 1970, Effects of a causeway on the chemistry of the brine in Great Salt Lake, Utah: Utah Geological Survey Water-Resources Bulletin 14, 52 p.

Pinder, G.F., and Cooper, H.H., Jr., 1970, A numerical technique for calculating the transient position of the saltwater front: Water Resources Research, v. 6, no. 3, p. 875–882.

ReMillard, M.D., 1982, Water resources data, Utah: U.S. Geological Survey Water-Data Report UT–82–1.

———1983, Water resources data, Utah: U.S. Geological Survey Water-Data Report UT–83–1.

———1984, Water resources data, Utah: U.S. Geological Survey Water-Data Report UT–84–1.

———1985, Water resources data, Utah: U.S. Geological Survey Water-Data Report UT–85–1.

———1986, Water resources data, Utah: U.S. Geological Survey Water-Data Report UT–86–1.

Sanford, W.E., and Konikow, L.F., 1985, A two-constituent solute-transport model for ground water having variable density: U.S. Geological Survey Water-Resources Investigations Report 85–4279, 88 p.

U.S. Geological Survey, 1980, Water resources data for Utah: U.S. Geological Survey Water-Data Report UT–80–1.

———1981, Water resources data for Utah: U.S. Geological Survey Water-Data Report UT–81–1.

Waddell, K.M., and Barton, J.D., 1980, Estimated inflow and evaporation for Great Salt Lake, Utah, 1931–76, with revised model for evaluating the effects of dikes on the water and salt balance of the lake: Utah Department of Natural Resources, Division of Water Resources Cooperative Investigation Report Number 20, 57 p.

Waddell, K.M., and Bolke, E.L., 1973, The effects of restricted circulation on the salt balance of Great Salt Lake, Utah: Utah Geological Survey Water-Resources Bulletin 18, 54 p.

Waddell, K.M., and Fields, F.K., 1977, Model for evaluating the effects of dikes on the water and salt balance of Great Salt Lake, Utah: Utah Geological Survey Water-Resources Bulletin 21, 54 p.

Wold, S.R., and Waddell, K.M., 1994, Salt budget for West Pond, Utah, April 1987 to June 1989: U.S. Geological Survey Water-Resources Investigations Report 93-4028, 20 p.

APPENDIXES

APPENDIX A. WATER BALANCE AND BOUNDARY CONDITIONS

The water balance between the south and north parts of the lake (fig. 5) consists of surface- and ground-water inflow (*SIS*, *GIS*, and *GIN*), precipitation on the lake surface (*PIS* and *PIN*), evaporation from the lake surface (*EOS* and *EON*), and flow through the causeway (*QS* and *QN*). Flow through the causeway, *QS* and *QN*, is interrelated with the boundary conditions, which consist of water-surface altitude and density of the south and north parts. Flow through the causeway, *QS* and *QN*, depends on freshwater inflow to the south part of the lake and on the conveyance properties of the causeway. Although total freshwater inflow to the lake consists of surface water, ground water, and precipitation, surface-water inflow makes up about 90 percent of the freshwater inflow that enters the south part of the lake. Flow through the causeway, *QS* and *QN*, is therefore essentially dependent on surface-water inflow to the south part.

The following equations, A1 and A2, relate the change in volume of the south and north parts during a given time interval, T, to surface- and ground-water inflow, precipitation, evaporation, and flow through the causeway. The methods used for estimating *SIS*, *PIS*, *PIN*, *GIS*, *GIN*, *EOS*, and *EON* are discussed later in this appendix. Flow through the causeway, *QS* and *QN*, consists of flow through the fill, culverts, and breach, which are discussed in Appendixes C, D, and E, respectively.

$$VS(I+1)=VS(I)+[SIS + PIS+GIS-EOS+QN -QS]\bullet T \quad \text{(A1)}$$

and

$$VN(I+1)=VN(I)+[PIN+GIN-EON+QS-QN]\bullet T, \quad \text{(A2)}$$

where

VS = volume of south part, in acre-feet;
VN = volume of north part, in acre-feet;
I+1 = next time step;
I = present time step;
SIS = surface-water inflow to south part, in acre-feet per day;
PIS = precipitation on south part, in acre-feet per day;
PIN = precipitation on north part, in acre-feet per day;
GIS = ground-water inflow to south part, in acre-feet per day;
GIN = ground-water inflow to north part, in acre-feet per day;
EOS = evaporation from south part, in acre-feet per day;
EON = evaporation from north part, in acre-feet per day;
QS = total flow from south-to-north through causeway, in acre-feet per day;
QN = total flow from north-to-south through causeway, in acre-feet per day; and
T = time interval, in days (in this case, 1.901 days).

Because the components of flow through the causeway, *QS* and *QN*, are interrelated with the differences in water-surface altitude and density, it was necessary to use a small time interval, *T*, in which the boundary conditions are held constant, to compute *QS* and *QN*. To minimize computation error associated with holding boundary conditions constant, the time interval can be no longer than 2 days. Total flow through the causeway, *QS* and *QN* (eqs A3 and A4), computed for each time interval, *T* (*T* equals 1.901 days), is the sum of the fill, culvert, and breach flow.

$$QS=1.9835(QSF+QSC+QSB) \quad \text{(A3)}$$

and

$$QN=1.9835(QNF+QNC+QNB), \quad \text{(A4)}$$

where

1.9835 acre-ft/d = 1 ft^3/s;
QSF = south-to-north flow through fill, in cubic feet per second;
QNF = north-to-south flow through fill, in cubic feet per second;
QSC = south-to-north flow through culverts, in cubic feet per second;
QNC = north-to-south flow through culverts, in cubic feet per second;
QSB = south-to-north flow through breach, in cubic feet per second; and
QNB = north-to-south flow through breach, in cubic feet per second.

Equations A1 and A2 were used to compute volume, *VS* and *VN*, for the next time step (*I+1*). Then, through the relations in table A1, water-surface altitudes for the south and north parts (*ES* and *EN*) can be computed as functions of the respective volumes, for the next time step.

The densities of the south and north parts of the lake also need to be computed for each time step. The relation of density to the water and salt balance of the lake is discussed in Appendix B.

Surface-Water Inflow

Streams and canals convey surface-water inflow to Great Salt Lake (*SIS* in eq A1). Most of the major streams and canals were gaged during 1980–86 (fig. 1), and measurements of monthly inflow from these sites were obtained from streamflow-gaging-station records (table A2). The other streams or canals that flow into Great Salt Lake were gaged during part of 1980–86 or prior to 1980. At the sites with incomplete records, empirical estimates of monthly inflow were derived using regression analysis between the sites with incomplete records and nearby sites with complete records (table A3). Monthly estimates of surface inflow

Table A1. Area and volume of Great Salt Lake, Utah, for selected water-surface altitudes

[Data from George Pyper, U.S. Geological Survey, written commun., 1986]

Water-surface altitude, in feet	South part		North part	
	Area, in acres	Volume, in acre-feet	Area, in acres	Volume, in acre-feet
4,171.0	123,900	116,100	48,010	42,100
4,171.5	131,700	180,000	53,820	67,500
4,172.0	139,500	247,800	59,640	95,900
4,172.5	147,300	319,500	65,450	127,200
4,173.0	155,100	395,100	71,260	161,400
4,173.5	162,900	474,600	77,080	198,500
4,174.0	170,700	558,000	82,890	238,500
4,174.5	178,600	645,300	88,700	281,400
4,175.0	186,400	736,500	94,520	327,200
4,175.5	194,200	831,600	100,300	375,900
4,176.0	202,000	930,600	106,100	427,500
4,176.5	209,800	1,033,500	112,000	482,000
4,177.0	217,600	1,140,300	117,800	539,400
4,177.5	225,400	1,251,000	123,600	599,700
4,178.0	233,200	1,365,600	129,400	662,900
4,178.5	241,000	1,484,100	135,200	729,000
4,179.0	248,800	1,606,500	141,000	798,000
4,179.5	256,700	1,732,900	146,800	869,900
4,180.0	264,400	1,863,200	152,600	944,700
4,180.5	269,600	1,996,700	155,600	1,021,800
4,181.0	274,700	2,132,800	158,700	1,100,400
4,181.5	279,800	2,271,400	161,700	1,180,500
4,182.0	284,900	2,412,600	164,700	1,262,100
4,182.5	290,000	2,556,300	167,700	1,345,200
4,183.0	295,100	2,702,600	170,700	1,429,800
4,183.5	300,200	2,851,400	173,800	1,515,900
4,184.0	305,400	3,002,800	176,800	1,603,600
4,184.5	310,500	3,156,800	179,800	1,692,800
4,185.0	315,600	3,313,300	182,800	1,783,500
4,185.5	320,700	3,472,400	185,800	1,875,700
4,186.0	325,800	3,634,000	188,900	1,969,400
4,186.5	330,900	3,798,200	191,900	2,064,600
4,187.0	336,000	3,964,900	194,900	2,161,300
4,187.5	341,200	4,134,200	197,900	2,259,500
4,188.0	346,300	4,306,100	200,900	2,359,200
4,188.5	351,400	4,480,500	204,000	2,460,400
4,189.0	356,500	4,657,500	207,000	2,563,200
4,189.5	361,600	4,837,000	210,000	2,667,500
4,190.0	366,700	5,019,100	213,000	2,773,300
4,190.5	372,800	5,204,000	217,100	2,880,800
4,191.0	378,900	5,391,900	221,200	2,990,400
4,191.5	384,900	5,582,900	225,300	3,102,000
4,192.0	391,000	5,776,900	229,400	3,215,700
4,192.5	397,100	5,973,900	233,500	3,331,400
4,193.0	403,200	6,174,000	237,700	3,449,200
4,193.5	414,300	6,378,400	247,500	3,570,500
4,194.0	425,400	6,588,300	257,300	3,696,700
4,194.5	436,600	6,803,800	267,200	3,827,800
4,195.0	447,800	7,024,900	277,000	3,963,800
4,195.5	467,900	7,253,800	287,800	4,105,000
4,196.0	488,000	7,492,800	298,600	4,251,600
4,196.5	508,100	7,741,800	309,400	4,403,600
4,197.0	528,200	8,000,900	320,300	4,561,000
4,197.5	548,400	8,270,000	331,100	4,723,800
4,198.0	568,500	8,549,200	341,900	4,892,000
4,198.5	588,600	8,838,500	352,700	5,065,600
4,199.0	608,700	9,137,800	363,500	5,244,600
4,199.5	628,800	9,447,200	374,400	5,429,100
4,200.0	648,900	9,766,600	385,200	5,619,000
4,200.5	691,020	10,108,700	392,800	5,813,500
4,201.0	704,720	10,457,600	400,300	6,011,800
4,201.5	718,310	10,813,400	407,900	6,213,900
4,202.0	731,910	11,176,000	415,500	6,419,800
4,202.5	745,600	11,545,400	423,100	6,629,500
4,203.0	759,190	11,921,600	430,700	6,843,000
4,203.5	772,890	12,304,600	438,300	7,060,300
4,204.0	786,470	12,694,500	445,900	7,281,400
4,204.5	798,480	13,090,800	453,500	7,506,300
4,205.0	810,580	13,493,100	461,100	7,735,000
4,205.5	821,080	13,901,000	467,200	7,967,100
4,206.0	831,590	14,314,200	473,300	8,202,200
4,206.5	842,190	14,732,700	479,400	8,440,400
4,207.0	852,700	15,156,400	485,600	8,681,700
4,207.5	863,300	15,585,400	491,700	8,926,000
4,208.0	874,010	16,019,700	497,800	9,173,400
4,208.5	910,110	16,465,700	503,900	9,423,800
4,209.0	946,210	16,929,800	510,000	9,677,300
4,209.5	968,520	17,408,500	513,600	9,933,200
4,210.0	977,720	17,895,100	517,300	10,190,900
4,210.5	986,830	18,386,200	520,900	10,450,500
4,211.0	996,030	18,881,900	524,500	10,711,900
4,211.5	1,005,140	19,382,200	528,100	10,975,100
4,212.0	1,014,340	19,887,100	531,900	11,240,100
4,212.5	1,021,530	20,396,100	551,300	11,510,800
4,213.0	1,028,830	20,908,700	572,700	11,791,700
4,213.5	1,036,120	21,424,900	596,500	12,083,900
4,214.0	1,043,410	21,944,800	617,800	12,387,500
4,214.5	1,050,710	22,468,300	637,000	12,701,200
4,215.0	1,058,000	22,995,500	825,300	13,311,100
4,215.5	1,065,300	23,526,300	884,500	13,738,300
4,216.0	1,072,490	24,060,800	945,700	14,195,700

(*SIS*) to Great Salt Lake for all streams and canals are listed in table A4.

The total annual surface-water inflow to the lake for 1980–86 ranged from 1,612,000 acre-ft in 1981 to 6,882,000 acre-ft in 1984 and averaged 4,627,000 acre-ft (table A5). The total annual surface-water inflow is distributed as follows: Bear River drainage system, 54 percent; Jordan River drainage system, 22 percent; Weber River drainage system, 15 percent; miscellaneous tributaries, 8 percent; and sewage inflow, 1 percent.

Bear River and Weber River Basins

Monthly flow into Great Salt Lake from the Bear River Basin was estimated from the record at Bear River Basin outflow across State Highway 83 near Corinne (site 10127110). Flow at this site was estimated from 3 continuous-recording gaging stations (sites 10126000, 10126180, and 10127100, fig. 1) and 46 culvert or bridge openings that cross State Highway 83. Station 10127110 was discontinued after September 1986; therefore, the monthly flow for October through December 1986 was estimated by multiplying a factor of 1.1 times monthly flow at Bear River near Corinne (site 10126000). The factor of 1.1 was used by Waddell and Barton (1980, p. 11) for the same site, and the ratio of monthly flow for October through December 1980–85 from the records at both sites was also 1.1. The monthly flow into Great Salt Lake from the Weber River Basin was estimated from the record at Weber River at Plain City (site 10141000).

Jordan River Basin

Flow from the Jordan River Basin was separated into a part that flows into Farmington Bay and a second part that flows into the south part of Great Salt Lake. Flow from the Jordan River and part of the flow from the Surplus Canal enters Farmington Bay. Flow from the Jordan River was estimated from the record at Jordan River at 500 North, at Salt Lake City (site 10172550, fig. 1). Water in the Surplus Canal flows to Farmington Bay or to the south part of Great Salt Lake via Goggin Drain (site 10172630). Total flow in the Surplus Canal was estimated from the record at Surplus Canal at Salt Lake City (site 10170500). The Surplus Canal flow to Farmington Bay was estimated by subtracting estimated monthly flow through Goggin Drain from the total. Records at Goggin Drain near Magna (site 10172630) were used to estimate flow from January 1980 to September 1984. Flow during the remaining period (October 1984 through December 1986) was estimated using a regression equation developed for monthly flow for 1977–84 at stations 10170500 and 10172630 (table A3).

Other Surface-Water Inflow

Monthly flow at Hooper Slough near Hooper (site 10141040) and Howard Slough at Hooper (site 10141400) was estimated from station records from October 1980 through September 1984. Monthly flow from October 1984 through December 1986 was estimated using regression equations with monthly flow at Black Slough near Brigham City (site 10127100) as the independent variable (table A3).

Table A2. Streamflow-gaging stations used to estimate monthly surface-water inflow to Great Salt Lake, Utah, 1980–86

Station number	Station name	Period of record during 1980–86
10127110	Bear River Basin outflow across State Highway 83 near Corinne	January 1980 to September 1986.
10141000	Weber River near Plain City	January 1980 to December 1986.
10141040	Hooper Slough near Hooper	January 1980 to September 1984.
10141400	Howard Slough at Hooper	Do.
10141500	Holmes Creek near Kaysville	Prior to January 1980.
10142000	Farmington Creek above diversions, near Farmington	Do.
10142500	Ricks Creek above diversions, near Centerville	Do.
10143000	Parrish Creek above diversions, near Centerville	Do.
10143500	Centerville Creek above diversions, near Centerville	Do.
10144000	Stone Creek above diversions, near Bountiful	Do.
10145000	Mill Creek at Mueller Park, near Bountiful	Do.
10170500	Surplus Canal at Salt Lake City	January 1980 to December 1986.
10170800	Surplus Canal at Cohen Flume, near Salt Lake City	Do.
10172550	Jordan River at 500 North, at Salt Lake City	January 1980 to September 1986.
10172630	Goggin Drain near Magna	January 1980 to September 1984.
10172640	Lee Creek near Magna	January 1980 to September 1982.
10172650	Kennecott Drain near Magna	January 1980 to September 1984.

Table A3. Statistical summary of regression estimates of monthly surface-water inflow to Great Salt Lake, Utah, 1980–86

Site estimated (dependent variable)	Site used for regression analysis (independent variable)	Number of months used for regression	Period being estimated, in months and years	Correlation coefficient	Average of dependent variable, in acre-feet	Standard error of estimate: In acre-feet	Standard error of estimate: As percentage of average of dependent variable
10141040	10127100	10	Jan.–Feb. 1985–86	0.597	749	232	31
		10	Mar.–Apr. 1985–86	.792	917	325	35
		15	May–July 1985–86	.257	1,440	310	22
		10	Aug.–Sept. 1985–86	.555	1,440	388	27
		12	Oct.–Dec. 1984–85	.671	798	311	39
10141400	10127100	10	Jan.–Feb. 1985–86	.691	1,440	415	29
		10	Mar.–Apr. 1985–86	.802	2,030	893	44
		10	May–June 1985–86	.274	2,140	475	22
		15	July–Sept. 1985–86	.850	2,420	455	19
		12	Oct.–Dec. 1984–85	.879	1,680	349	21
[1]10141500	10172500	[2]16	Jan.–Dec. 1980–86	.911	2,670	374	14
[1]10142000	10172500	[2]26	Jan.–Dec. 1980–86	.923	9,610	1,440	15
[1]10142500	10172500	[2]16	Jan.–Dec. 1980–86	.859	1,610	370	23
[1]10143000	10172500	[2]19	Jan.–Dec. 1980–86	.952	1,140	160	14
[1]10143500	10172500	[2]31	Jan.–Dec. 1980–86	.969	2,190	219	10
[1]10144000	10172500	[2]16	Jan.–Dec. 1980–86	.943	2,290	389	17
[1]10145000	10172500	[2]18	Jan.–Dec. 1980–86	.951	4,660	699	15
[3]10172630	10170500	46	Oct.–Dec. 1984–86	.984	18,210	2,648	15
[4]10172630	10170500	19	Oct.–Dec. 1984–86	.769	66,390	8,630	13

[1] Annual total. Annual flow was estimated using regression equation. Monthly flows were estimated using average monthly fractions of the annual flows based on station records.
[2] Number of years used for regression analysis.
[3] Based on monthly flows less than 70,000 acre-feet at station 10170500.
[4] Based on monthly flows greater than or equal to 70,000 acre-feet at station 10170500.

Monthly flow at Lee Creek near Magna (site 10172640) for January 1980 through September 1982 and Kennecott Drain near Magna (site 10172650) for January 1980 through September 1984 was estimated from station records. Monthly flow at Lee Creek for October 1982 through December 1986 was estimated as the average monthly flow for the station record from January 1972 through September 1982. Monthly flow at Kennecott Drain for October 1984 to December 1986 was estimated as average monthly flow for the station record from January 1972 through September 1984.

Monthly flow for the seven perennial streams in Davis County (sites 10141500 to 10145000, table A4) between the Weber River and the Jordan River was estimated using regression equations for each stream that relate annual flow at City Creek (site 10172500) to annual flow at each stream (David Clark, U.S. Geological Survey, written commun., 1987) (table A3). Estimated annual flow for 1980–86 was divided into monthly flow using the average fraction (monthly flow/annual flow) for each month determined using the station records for the seven stations.

Most or all water in the Davis County streams is diverted for irrigation or municipal use at the point where the streams emerge from the canyons onto the benches of the East Shore area. Monthly flow used in the water balance is the estimated flow at the mouths of the canyons, before any diversions. Although most flow at the canyon mouths usually does not reach Great Salt Lake, it can be used as an index of the relative amount of irrigation return flow.

The Salt Lake City water-reclamation plant releases water into Great Salt Lake, and estimates of that flow were obtained from records at the plant. Estimates of the amount of flow released into Great Salt Lake from Willard Bay reservoir were obtained from the Weber Basin Water Conservancy District.

Table A4. Estimated monthly contributions of surface-water inflow to Great Salt Lake, Utah, 1980–86

[In acre-feet; all estimates based on regression equations or averages are reported to two significant figures except regression estimates for Goggin Drain, which are reported to four significant figures to facilitate computations]

Year	January	February	March	April	May	June	July	August	September	October	November	December	Total
10127110, Bear River Basin outflow across State Highway 83 near Corinne													
1980	168,500	145,700	125,400	152,900	310,600	311,600	77,860	70,140	120,200	133,700	135,500	131,200	1,883,000
1981	117,800	75,140	76,580	76,640	90,700	89,710	24,160	22,160	28,700	70,030	65,080	76,730	813,400
1982	69,300	68,200	192,700	216,300	331,800	192,800	65,490	83,450	147,400	211,200	196,100	207,000	1,982,000
1983	202,000	183,500	282,300	241,400	402,700	490,700	278,400	210,800	201,400	282,900	275,800	296,000	3,348,000
1984	239,400	220,100	345,100	415,400	605,900	568,000	226,300	201,500	222,900	269,800	283,200	243,200	3,841,000
1985	183,100	155,800	244,200	443,900	231,800	58,580	89,800	95,390	124,000	139,100	156,100	161,100	2,083,000
1986	152,900	353,300	388,100	435,000	503,000	412,400	226,400	157,300	210,700	220,000	230,000	210,000	3,499,000
10141000, Weber River near Plain City													
1980	16,390	34,580	43,850	129,600	166,800	79,080	6,440	3,510	9,640	13,850	18,650	22,860	545,200
1981	6,990	5,370	6,950	10,560	35,180	38,430	4,350	4,110	5,050	9,590	8,390	8,900	143,900
1982	8,470	23,470	82,430	162,700	124,700	64,260	18,340	6,950	22,050	37,180	44,540	48,990	644,100
1983	38,880	36,110	68,020	126,600	220,600	251,900	37,030	25,450	57,590	42,450	38,740	115,800	1,059,000
1984	104,000	29,210	51,570	118,300	252,800	169,300	18,050	12,320	30,790	59,530	14,920	32,890	893,700
1985	84,740	68,420	71,110	153,200	134,900	27,120	7,410	6,790	15,640	20,390	15,150	26,050	630,900
1986	29,200	133,200	215,300	216,500	245,300	82,910	10,790	10,520	27,960	52,840	38,680	38,390	1,102,000
10141040, Hooper Slough near Hooper													
1980	916	1,250	896	427	896	1,430	1,370	1,320	1,540	1,110	498	543	12,200
1981	469	496	652	418	1,470	1,070	1,240	847	831	901	502	454	9,350
1982	523	563	1,320	526	1,430	1,620	2,070	1,490	1,730	1,090	555	747	13,700
1983	655	1,060	1,300	823	1,810	1,610	1,510	2,340	1,460	711	620	1,850	15,700
1984	943	619	2,020	790	1,720	1,340	1,030	1,630	1,200	810	990	870	14,000
1985	790	680	1,300	900	1,400	1,400	1,400	1,200	1,600	680	740	780	13,000
1986	680	1,400	1,400	1,400	1,500	1,400	1,400	1,300	1,500	1,100	620	900	15,000
10141400, Howard Slough at Hooper													
1980	1,350	1,630	1,460	549	2,060	2,050	1,520	1,320	1,830	1,540	767	942	17,000
1981	801	901	1,380	1,100	2,130	1,650	1,470	1,460	1,860	1,930	1,000	1,010	16,700
1982	976	965	2,040	1,050	2,770	1,630	2,920	2,240	3,110	2,350	1,710	2,340	24,100
1983	1,860	2,470	3,340	1,800	2,360	2,150	2,440	4,040	3,430	1,910	1,530	3,110	30,400
1984	1,900	1,510	5,390	2,200	1,640	3,000	2,670	2,870	3,120	1,700	2,100	1,800	30,000
1985	1,500	1,300	3,200	2,000	2,000	2,000	1,900	1,900	2,900	1,400	1,500	1,600	23,000
1986	1,300	3,000	3,400	3,300	2,200	2,000	2,200	2,100	2,900	2,100	1,900	2,100	28,000
Combined flow at streamflow-gaging stations 10141500, 10142000, 10142500, 10143000, 10143500, 10144000, and 10145000													
1980	750	760	1,200	3,600	8,800	4,800	1,500	840	660	740	750	770	25,000
1981	590	600	930	2,800	6,900	3,800	1,200	660	520	590	590	600	20,000
1982	1,200	1,200	1,900	5,700	14,000	7,600	2,400	1,300	1,000	1,200	1,200	1,200	40,000
1983	2,400	2,400	3,800	11,000	28,000	16,000	4,800	2,700	2,100	2,400	2,400	2,400	80,000
1984	1,900	2,000	3,000	9,200	23,000	12,000	3,900	2,200	1,700	1,900	1,900	2,000	65,000
1985	1,200	1,200	1,800	5,600	14,000	7,500	2,400	1,300	1,000	1,200	1,200	1,200	40,000
1986	1,800	1,800	2,800	8,600	21,000	12,000	3,600	2,000	1,600	1,800	1,800	1,800	61,000
10170800, Surplus Canal at Cohen Flume, near Salt Lake City													
1980	7,150	10,360	8,890	8,260	12,380	16,370	12,310	11,670	12,210	11,300	12,060	14,190	137,200
1981	14,120	12,600	13,220	11,660	15,430	15,360	11,750	11,624	13,160	12,990	13,110	14,140	159,200
1982	12,470	11,360	11,230	13,590	17,270	19,870	18,970	11,720	19,180	24,130	18,020	15,500	193,300
1983	9,320	17,970	24,830	29,060	49,240	81,590	53,900	32,860	25,370	28,940	34,220	39,760	427,100
1984	39,950	36,960	40,840	54,080	101,600	125,200	71,200	45,070	29,710	31,910	32,790	33,880	643,200
1985	27,550	22,460	36,060	40,910	48,800	27,070	15,750	13,650	14,110	15,440	16,360	17,500	295,700
1986	17,480	30,430	16,620	84,170	101,800	85,290	39,960	25,370	26,120	16,950	20,680	30,190	495,100

Table A4. Estimated monthly contributions of surface-water inflow to Great Salt Lake, Utah, 1980–86—Continued

Year	January	February	March	April	May	June	July	August	September	October	November	December	Total
10172550, Jordan River at 500 North, at Salt Lake City													
1980	7,980	9,240	11,840	12,960	16,370	13,900	11,920	11,770	11,260	11,330	11,130	11,820	141,500
1981	12,110	10,850	12,220	11,510	13,180	11,360	11,440	11,460	10,920	11,250	9,230	10,620	136,200
1982	11,090	10,930	13,300	16,910	20,330	18,730	15,230	12,360	14,410	13,610	13,080	14,300	174,300
1983	15,550	12,090	17,430	19,860	29,130	36,960	20,990	18,810	15,220	15,400	15,020	16,430	232,900
1984	14,590	11,960	11,530	19,630	32,680	28,800	21,220	13,810	12,710	18,300	16,340	13,890	215,500
1985	19,270	16,240	13,850	19,840	23,780	20,010	16,850	15,620	18,010	16,770	15,970	15,850	212,100
1986	17,010	18,510	13,310	18,460	26,740	21,640	16,370	14,770	14,630	14,930	11,370	11,270	199,000
10172630, Goggin Drain near Magna													
1980	2,140	12,440	24,110	25,020	32,000	29,750	10,230	816	1,020	8,320	13,330	17,880	177,100
1981	19,510	19,140	24,500	19,190	14,560	14,240	1,100	856	658	2,860	2,340	5,520	124,500
1982	10,070	13,840	27,920	32,840	41,400	40,360	19,860	3,840	13,690	42,340	41,110	48,120	335,400
1983	56,690	46,850	55,220	57,840	74,660	90,710	78,800	68,640	53,950	61,610	61,960	67,240	774,200
1984	71,150	66,840	74,860	76,820	84,240	71,110	51,700	48,450	45,560	58,120	58,570	59,110	766,500
1985	55,840	52,990	60,180	62,490	66,000	55,580	36,940	19,380	23,280	34,340	42,070	51,590	560,700
1986	51,370	57,360	44,220	79,430	85,180	79,810	62,040	54,640	55,060	47,000	51,950	57,240	725,300
10172640, Lee Creek near Magna													
1980	167	267	274	171	209	61	181	311	377	255	87	117	2,480
1981	89	77	188	136	244	69	103	209	173	460	315	288	2,350
1982	297	210	336	379	172	77	188	203	992	260	180	180	3,500
1983	250	280	310	280	200	130	190	250	370	260	180	180	2,900
1984	250	280	310	280	200	130	190	250	370	260	180	180	2,900
1985	250	280	310	280	200	130	190	250	370	260	180	180	2,900
1986	250	280	310	280	200	130	190	250	370	260	180	180	2,900
10172650, Kennecott Drain near Magna													
1980	5,930	6,460	5,170	4,050	7,010	4,210	6,460	7,020	5,900	5,050	4,640	5,610	67,500
1981	5,290	4,600	4,890	3,730	5,190	3,300	3,330	3,370	4,830	7,380	5,650	5,210	56,800
1982	5,960	4,900	5,170	4,930	5,220	3,980	5,210	5,180	8,260	6,580	6,140	6,670	68,200
1983	6,540	5,290	8,080	7,440	7,920	5,420	5,800	7,550	7,180	8,090	8,450	8,980	86,700
1984	8,290	7,520	7,780	6,170	5,330	6,550	8,290	8,940	7,570	6,600	6,400	6,300	86,000
1985	6,000	6,100	6,600	5,400	6,300	5,100	5,600	6,400	6,800	6,600	6,400	6,300	74,000
1986	6,000	6,100	6,600	5,400	6,300	5,100	5,600	6,400	6,800	6,600	6,400	6,300	74,000
Outflow from Salt Lake City water-reclamation plant[1]													
1980	3,980	4,330	4,220	3,870	4,240	3,350	4,380	4,200	4,390	4,600	4,300	4,340	50,200
1981	3,460	3,260	3,830	3,860	4,520	4,200	4,200	4,150	3,660	4,060	3,770	3,890	46,900
1982	4,020	3,670	4,500	4,710	4,590	4,160	4,430	4,280	5,000	4,910	4,300	3,770	52,300
1983	3,610	3,630	3,910	3,860	3,960	3,780	3,230	3,630	3,520	3,600	3,380	3,300	43,400
1984	3,660	3,470	4,510	4,610	4,330	3,960	3,730	3,630	3,330	4,900	4,820	4,300	49,200
1985	4,300	4,300	4,300	3,920	4,280	3,850	3,840	3,690	3,490	3,470	3,360	3,550	46,400
1986	3,580	3,570	4,140	3,840	3,560	3,830	3,570	3,830	3,600	3,500	3,420	3,190	43,600
Outflow from Willard Bay reservoir[2]													
1980	1,090	13,280	25,920	46,070	32,640	19,080	5,880	5,720	3,420	3,620	1,430	1,330	159,000
1981	1,410	1,270	1,890	11,870	1,750	7,540	21,690	19,930	11,180	850	2,320	980	82,700
1982	12,030	9,910	21,750	48,120	33,230	5,710	2,480	6,350	1,780	40,550	32,640	700	215,000
1983	5,230	2,680	27,970	57,070	54,220	49,190	7,930	5,010	5,610	3,280	2,180	33,320	254,000
1984	56,790	32,000	34,040	32,190	49,630	21,620	3,480	6,950	12,510	10,080	10,110	6,200	276,000
1985	3,400	400	100	48,090	20,100	4,490	5,150	6,130	3,570	2,700	320	4,930	99,400
1986	9,740	18,120	47,870	55,250	58,200	44,310	8,520	6,130	3,160	1,580	990	400	254,000

[1] Records from Salt Lake City water-reclamation plant.

[2] Records from Weber Basin Water Conservancy District, Layton, Utah.

Table A5. Estimated monthly surface-water inflow to Great Salt Lake, Utah, 1980–86

[In acre-feet]

Year	January	February	March	April	May	June	July	August	September	October	November	December	Total
1980	216,300	240,300	253,200	387,500	594,000	485,700	140,100	118,600	172,400	195,400	203,100	211,600	3,218,000
1981	182,600	134,300	147,200	153,500	191,300	190,700	86,030	80,830	81,540	122,900	112,300	128,300	1,612,000
1982	136,400	149,200	364,600	507,800	596,900	360,800	157,600	139,400	238,600	385,400	359,600	349,500	3,746,000
1983	343,000	314,300	496,500	557,000	874,800	1,030,000	495,000	382,100	377,200	451,600	444,500	588,400	6,354,000
1984	542,800	412,500	581,000	739,700	1,163,000	1,011,000	411,800	347,600	371,500	463,900	432,300	404,600	6,882,000
1985	387,900	330,200	443,000	786,500	553,600	212,800	187,200	171,700	214,800	242,400	259,400	290,600	4,080,000
1986	291,300	627,100	744,100	911,600	1,055,000	750,800	380,600	284,600	354,400	368,700	368,000	362,000	6,498,000

Ground-Water Inflow

Ground-water inflow (*GIN* + *GIS*, eqs A1 and A2) to Great Salt Lake was estimated to be 75,000 acre-ft/yr. The estimates were based on the previous water-balance study of Great Salt Lake (Waddell and Fields, 1977, p. 22). This inflow was subdivided for Farmington Bay, Bear River Bay, the shoreline extending from Bear River Bay to Syracuse, and the south and north parts of the lake (table A6).

Precipitation

Inflow to Great Salt Lake by precipitation directly on the water surface (*PIN* + *PIS*, eqs A1 and A2) was estimated using a method similar to that of Waddell and Fields (1977, p. 6 and 7) as follows:

1. Waddell and Fields (1977) compiled average annual precipitation data for 1931–73 for 68 sites in a large area surrounding the lake. A multiple regression equation estimating average annual precipitation as a function of latitude, longitude, and altitude was derived. Using the equation, lines of equal average annual precipitation during 1931–73 were drawn for the lake (Waddell and Fields, 1977, fig. 3). The relative distribution of average annual precipitation for 1980–86 was assumed to be the same as for 1931–73.
2. The surface area of the lake varies with water-surface altitude, and precipitation varies areally across the lake; thus, inflow from precipitation on the lake varies according to water-surface altitude. Waddell and Fields (1977, p. 6) derived values of average annual precipitation for four altitudes ranging from 4,195 to 4,205 ft for the south and north parts of the lake. For this study, average annual precipitation (1931–73) also was determined for a water-surface altitude of 4,212 ft (table A7).
3. Waddell and Fields (1977, p. 6 and 7) estimated total precipitation on the lake for each year of the study as a fraction of average annual precipitation. Precipitation for each month of each year was estimated as a fraction of the annual total. The fraction used for each month (Waddell and Fields, 1977, fig. 4) was an average value and was the same for all years of the study.

Table A6. Estimated ground-water inflow to Great Salt Lake, Utah, 1980–86

(From Waddell and Fields, 1977, table 8)

Area of inflow	Monthly inflow, in acre-feet	Annual inflow, in acre-feet
Farmington Bay	2,290	
Bear River Bay	1,250	
Bear River Bay to Syracuse	1,000	
South part of Great Salt Lake	870	
North part of Great Salt Lake	830	
Total monthly	6,240	
Total annual (rounded)		75,000

The monthly distribution of precipitation for 1980–86 was substantially different from the average distribution used by Waddell and Fields (1977, p. 7, fig. 4). In the water balance, monthly precipitation for the lake therefore was estimated using an average ratio index (*PRT*) determined from measured monthly precipitation at three sites around the lake (Salt Lake City Airport, Ogden Sugar Factory, and Tooele; fig. 1).

The derivation of the equation used in the water balance to estimate monthly precipitation for the south part of the lake is shown in the following equations. The *PRT*, based on data from three precipitation stations around the lake, is the average of a ratio from each station. The ratio is the measured monthly precipitation at a given precipitation station to the average annual precipitation (1931–73) at the same precipitation station.

$$PRT = [(PMSLC/PAASLC)+(PMOGSF/PAAOGSF)+(PMTOOL/PAATOOL)]/3, \qquad (A5)$$

where

PRT = average ratio index for precipitation;
$PMSLC$ = measured monthly precipitation at Salt Lake City Airport, in inches;
$PAASLC$ = average annual precipitation (1931–73) at Salt Lake City Airport, in inches;
$PMOGSF$ = measured monthly precipitation at Ogden Sugar Factory, in inches;
$PAAOGSF$ = average annual precipitation (1931–73) at Ogden Sugar Factory, in inches;
$PMTOOL$ = measured monthly precipitation at Tooele, in inches; and
$PAATOOL$ = average annual precipitation (1931–73) at Tooele, in inches.

The equations used in the water balance to estimate monthly precipitation for the south and north parts, in acre-feet per day are

$$PIS=PRT \bullet PAAS \bullet CF \bullet AS/CM \qquad (A6)$$

and

$$PIN=PRT \bullet PAAN \bullet CF \bullet AN/CM, \qquad (A7)$$

where

PIS = precipitation on south part, in acre-feet per day;
PIN = precipitation on north part, in acre-feet per day;
PRT = average ratio index (see eq A5);
$PAAS$ = average annual precipitation (1931–73) for the south part of the lake, interpolated from table A7 using water-surface altitude, in inches;
$PAAN$ = average annual precipitation (1931–73) for the north part of the lake, interpolated from table A7 using water-surface altitude, in inches;
CF = 1/12 (conversion between inches and feet), in feet per inch;
AS = area of south part for time step I, estimated from south part water-surface altitude (table 2), in acres;
AN = area of north part for time step I, estimated from north part water-surface altitude (table 2) in acres; and
CM = time interval (in this case, 365/12 days, about 1 month), in days.

Evaporation

Evaporation ($EON + EOS$, eqs A1 and A2) from Great Salt Lake was estimated using the same method employed by Waddell and Fields (1977, p. 7–11). The pan-evaporation records for 1980–86 were examined for Bear River Refuge, Saltair, and Utah Lake at Lehi (fig. 1). The fractions of annual evaporation for each month during 1980–86 were similar to the average fractions for 1931–73 used by Waddell and Fields (1977, fig. 4); therefore, the same monthly fractions used by Waddell and Fields were used in this water balance. The pan-evaporation values for June through September 1980–86 at all three sites (table A8) were less than the mean values for 1931–73 estimated by Waddell and Fields (1977, table 12). The average fractions of June through September pan evaporation for the three stations in table A8 were therefore used in the water balance to estimate annual evaporation from the lake.

The following equations are used in the water balance to estimate monthly evaporation for the south and north parts, respectively, in acre-feet:

$$EOS = EAAS \bullet EAI \bullet EMI \bullet CF \bullet AS \bullet SCFS/CM \qquad (A8)$$

and

$$EON = EAAN \bullet EAI \bullet EMI \bullet CF \bullet AN \bullet SCFN/CM, \qquad (A9)$$

where

EOS = evaporation from south part, in acre-feet per day;

Table A7. Average annual 1980–86 precipitation and evaporation of freshwater for selected water-surface altitudes of Great Salt Lake, Utah

[PAAS: Average annual precipitation for south part; EAAS: Average annual freshwater evaporation from south part; PAAN: Average annual precipitation for north part; EAAN: Average annual freshwater evaporation from north part]

Area of Great Salt Lake	Water-surface altitude, in feet	Precipitation[1], in inches	Evaporation[2], in inches
		PAAS	EAAS
South part	4,212	12.60	55.07
	4,205	12.98	55.98
	4,199	13.46	56.25
	4,196	13.70	56.39
	4,195	13.74	56.41
		PAAN	EAAN
North part	4,212	10.67	63.14
	4,205	10.66	62.72
	4,199	10.80	62.09
	4,196	11.08	61.48
	4,195	11.13	61.32

[1]Average annual precipitation for 1980–86 was assumed to be the same as for 1931—73.

[2]Average annual evaporation for 1980–86 was assumed to be the same as for 1931—73, except for June through September evaporation.

Table A8. June through September pan evaporation and percentage of mean pan evaporation (1931–73) for three sites near Great Salt Lake, Utah, 1980–86

[Mean pan evaporation for 1931–73 was estimated by Waddell and Fields (1977, p. 7 and table 12)]

Year	Bear River Bird Refuge		Saltair		Utah Lake at Lehi	
	Evaporation, in inches	Percentage of mean	Evaporation, in inches	Percentage of mean	Evaporation, in inches	Percentage of mean
1980	31.63	84	47.53	88	35.29	97
1981	34.60	92	48.39	90	34.31	94
1982	31.95	85	43.97	82	29.62	82
1983	29.46	78	44.4	83	[1]29.22	80
1984	—	—	38.58	72	24.72	68
1985	—	—	45.32	84	34.56	95
1986	—	—	44.51	83	35.10	97

[1]Evaporation data for June 1983 at Utah Lake in Lehi is missing, and the June value was estimated using an average ratio of June to July values for the station during 1980–82 and 1984–86.

EON = evaporation from north part, in acre-feet per day;

EAAS = average annual freshwater evaporation from south part, interpolated from table A7 using water-surface altitude, in inches;

EAAN = average annual freshwater evaporation from north part, interpolated from table A7 using water-surface altitude, in inches;

EAI = fraction of average annual evaporation (average of the three sites in table A8);

EMI = fraction of average monthly evaporation, from Waddell and Fields (1977, fig. 4);

CF = 1/12 (conversion between inches and feet), in feet per inch;

AS = area of south part for time step *I*, estimated from south part water-surface altitude (table 2), in acres;

AN = area of north part for time step *I*, estimated from north part water-surface altitude (table A1), in acres;

SCFS = salinity correction factor for evaporation rate in the south part;

SCFN = salinity correction factor for evaporation rate in the north part; and

CM = time interval (in this case, 365/12 days, about 1 month), in days.

The correction factors for salinity (*SCFS* and *SCFN* in eqs A8 and A9, respectively) were estimated for each month from curves of evaporation as a function of dissolved-solids concentration. The curves for the south and north parts of the lake were drawn based on measurements of dissolved-solids concentration made about four times a year. The measurements were plotted on a graph, a smooth curve was drawn through the measurements, then individual monthly values of salinity were obtained from the fitted curves. The salinity correction factors were estimated with the following equations using data from Adams (1934) and developed by Waddell and Bolke (1973):

$$SCFS=(1-0.778\ CS/\rho s) \qquad (A10)$$

$$SCFN=(1-0.778\ CN/\rho n) \qquad (A11)$$

where

SCFS = salinity correction factor for evaporation rate in the south part;

SCFN = salinity correction factor for evaporation rate in the north part;

CS = dissolved-solids concentration in the south part, in grams per milliliter;

CN = dissolved-solids concentration in the north part, in grams per milliliter;

ρs = density of brine in south part at any temperature, in grams per milliliter; and

ρn = density of brine in north part at any temperature, in grams per milliliter.

Calibration

A water balance between the south and north parts of Great Salt Lake was calibrated for 1980–86 using monthly surface- and ground-water inflow (*SIS*, *GIS*, and *GIN*), precipitation (*PIS* and *PIN*), evaporation (*EOS* and *EON*), and net flow through the causeway (*QS* - *QN*). The purpose of the water balance is to test and modify the independent estimates of monthly inflow, precipitation, and evaporation so that the simulated water-surface altitudes for the south part reasonably match the measured water-surface altitudes. By estimating the volume of the north part (*VN*) from the measured water-surface altitude of the north part (*EN*), all the error in simulating the water balance (from both the south and north parts) appeared as error in simulating the volume of the south part (*VS*) and water-surface altitude of the south part (*ES*).

The change in volume of the north part (ΔVN) was estimated from the difference in measured water-surface altitude in the north part from the beginning to the end of the time interval, using the water-surface altitude-area-volume relations (see Appendix A, section entitled "Water balance and boundary conditions"). After ΔVN and the ground-water inflow (*GIN*), precipitation (*PIN*), and evaporation (*EON*) for the north part were estimated, equation A12 (rearranged from eq A2) was used to solve for net flow through the causeway (*QS-QN*), in acre-feet per day.

$$(QS-QN)=\Delta VN/CM-(PIN+GIN-EON) \qquad \text{(A12)}$$

where

- *QS* = total flow from south-to-north through causeway, in acre-feet per day;
- *QN* = total flow from north-to-south through causeway, in acre-feet per day;
- ΔVN = $VN(I+1)-VN(I)$ = change in volume of north part, in acre-feet;
- *CM* = time interval (in this case, 365/12 days, about 1 month), in days;
- *PIN* = precipitation on north part, in acre-feet per day;
- *GIN* = ground-water inflow to north part, in acre-feet per day; and
- *EON* = evaporation from north part, in acre-feet per day.

Net flow through the causeway (*QS-QN*) is the difference between total flow through the causeway from south to north (*QS*) and total flow through the causeway from north to south (*QN*). To simulate the water-surface altitude of the south part, only the difference between *QS* and *QN* needs to be known, not each individual flow. For convenience, net flow through the causeway was always assumed to be a positive value of south-to-north flow. After the net flow through the causeway (*QS* - *QN*) and the surface- and ground-water inflow (*SIS* and GIS), precipitation (*PIS*), and evaporation (*EOS*) for the south part were estimated, equation A13 (rearranged from eq A1) was used to compute the change in volume of the south part, ΔVS, in acre-feet, as follows:

$$\Delta VS=[SIS+PIS+GIS-EOS-(QS-QN)]\bullet CM \qquad \text{(A13)}$$

where

- ΔVS = $VS(I+1)-VS(I)$ = change in volume of south part, in acre-feet;
- *SIS* = surface-water inflow to south part, in acre-feet per day;
- *PIS* = precipitation on south part, in acre-feet per day;
- *GIS* = ground-water inflow to south part, in acre-feet per day; and
- *EOS* = evaporation from south part, in acre-feet per day.

The initial estimates of inflow and outflow for the water balance resulted in simulated water-surface altitudes that were lower than measured water-surface altitudes by a maximum of about 3 ft (fig. A1); thus, the water balance included either too little inflow or too much outflow (evaporation). Calibration of the water balance was achieved by multiplying a constant factor times both monthly net inflow and evaporation and by comparing simulated with measured water-surface altitudes of the south part. The range of factors used in the calibration was 1.00 to 1.10 for precipitation, 1.00 to 1.10 for surface- and ground-water inflow, and 0.90 to 1.05 for evaporation. Many combinations of factors were tried and the best match of measured and simulated water-surface altitudes was obtained by multiplying all inflow (precipitation and surface- and ground-water inflow) times 1.07, and keeping evaporation the same as the original estimate. These simulated water-surface altitudes reasonably compared with measured water-surface altitudes in all years except 1984. For 1984, the initial estimate of evaporation of 70 percent of mean annual evaporation (1931–73) was changed to 80 percent for a better match. Simulated and measured water-surface altitudes for the calibrated water balance are shown in figure A1.

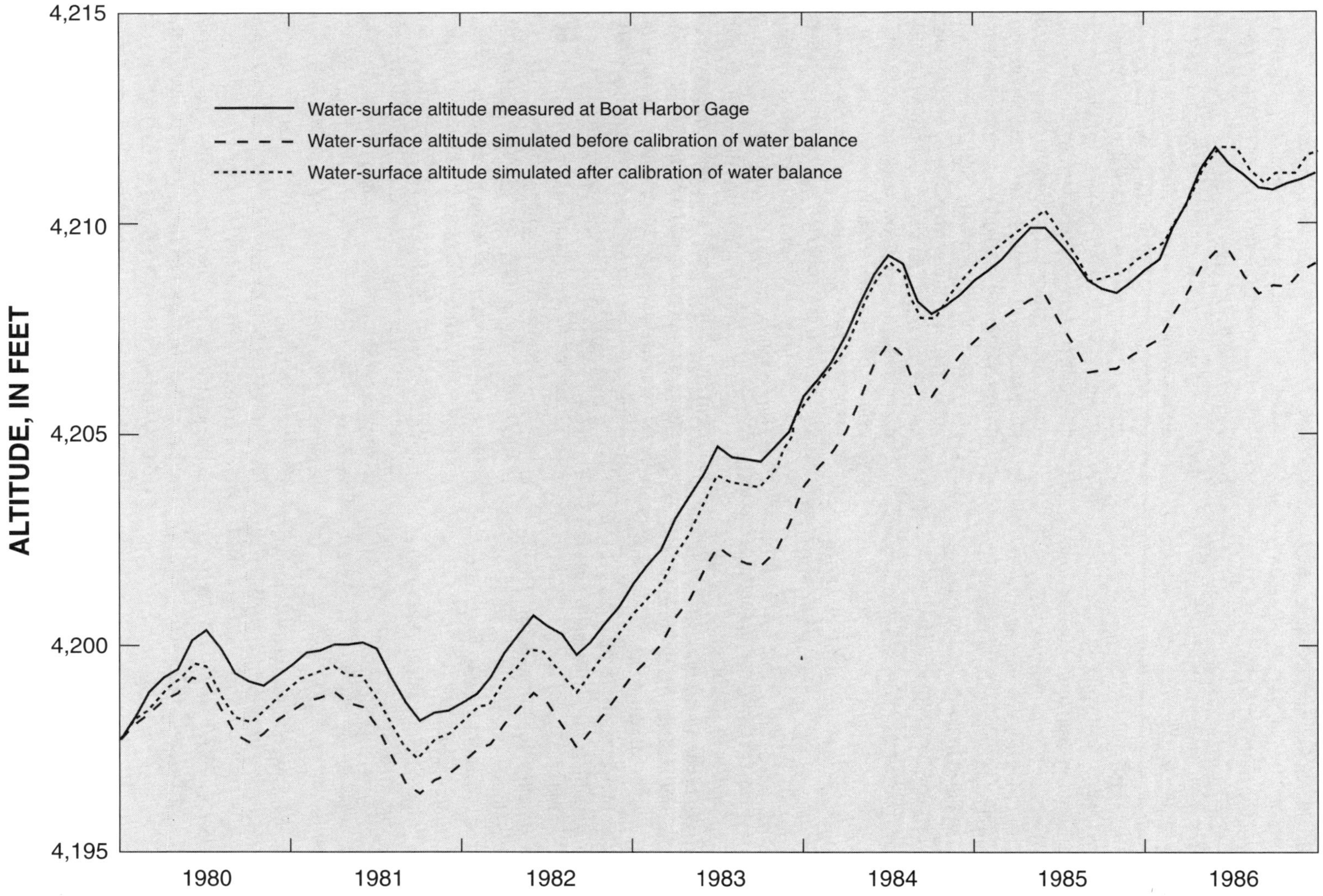

Figure A1. Simulated and measured water-surface altitude of the south part of Great Salt Lake, Utah, before and after calibration of water balance, 1980–86.

APPENDIX B. SALT BALANCE

A schematic diagram of the salt balance for the south and north parts of Great Salt Lake is shown in figure 7. Dissolved-solids concentration, density, and density difference were computed from the salt balance using the same procedure as in Waddell and Bolke (1973, p. 34). Total salt load in the lake (*LT*) for 1980–86 is represented by:

$$LT = LS + LSL + LSP + LN + LNP \qquad \text{(B1)}$$

where

LT = total salt load in lake, in tons;
LS = dissolved salt load in south part, in tons;
LSL = dissolved salt load in deep brine layer in south part, in tons;
LSP = precipitated salt load in south part, in tons;
LN = dissolved salt load in north part, in tons; and
LNP = precipitated salt load in north part, in tons.

The dissolved salt load in the south part (*LS*) at any time step *(I)* can be represented by

$$LS(I)=LS(I\text{-}1)\text{-}QS(LS(I\text{-}1))T/VS(I\text{-}1)+ QN(LN(I\text{-}1))T/VN(I\text{-}1)+LSD(I)\text{-}LSP(I) \qquad \text{(B2)}$$

where

$LS(I)$ = dissolved salt load in south part at time step *I*, in tons;
I = present time step;
$LS(I\text{-}1)$ = initial dissolved salt load in south part, in tons;
$I\text{-}1$ = previous time step;
$QS(LS(I\text{-}1))T/VS(I\text{-}1)$ = outflow load from south part, in tons;
$QN(LN(I\text{-}1))T/VN(I\text{-}1)$ = inflow load from north part, in tons;
$LSD(I)$ = redissolved salt in south part, in tons; and
$LSP(I)$ = precipitated salt load in south part, in tons.

Salt re-solution *LSD(I)* and salt precipitation *LSP(I)* cannot occur in the same time step.

Similarly, the dissolved load in the north part (*LN*), in tons, can be represented by

$$LN(I) = LN(I\text{-}1) + QS(LS(I\text{-}1))T/VS(I\text{-}1) \text{ - } QN(LN(I\text{-}1))T/VN(I\text{-}1) + LND(I) \text{ - } LNP(I) \qquad \text{(B3)}$$

where

$LN(I)$ = dissolved salt load in north part at time step *I*;
$LN(I\text{-}1)$ = initial dissolved salt load in north part;
$QS(LS(I\text{-}1))T/VS(I\text{-}1)$ = inflow load from south part;
$QN(LN(I\text{-}1))T/VN(I\text{-}1)$ = outflow load from north part;
$LND(I)$ = redissolved salt in north part; and
$LNP(I)$ = precipitated salt load in north part.

Salt Precipitation and Re-solution

The following is a synopsis of a description of the lake dynamics governing salt precipitation and re-solution by Waddell and Bolke (1973):

> Most salt precipitation in the north part occurs during summer and fall when the water surface is falling, and re-solution generally occurs during winter and spring when the water surface is rising. When the water surface is falling, water loss from evaporation in the north part may exceed the net gain of water to the north part; therefore, the concentration of dissolved solids increases in the north part, and if saturation concentration is attained (355 g/L), sodium chloride may precipitate. When the water surface is rising, the net gain of water in the north part may exceed water loss from evaporation and the concentrations in the north part may be diluted below saturation concentration. If dilution occurs, then conditions are conducive to re-solution of salt. Whether there is a net increase of salt precipitation or re-solution in the north part depends upon the magnitude of salt gain relative to net water gain from flow through the causeway.

Over periods of a few years, the addition of dissolved salts from inflow into Great Salt Lake is negligible compared with the total salt load of the lake. In the lake, salt can be either dissolved as *LS*, *LN*, and *LSL* (fig. 7), or be present in its precipitated form as *LSP* and *LNP*. During 1980–86, salt precipitated (*LNP*) only in the north part.

Estimates of precipitation and re-solution of sodium chloride in Great Salt Lake during 1980–86 were based on results of water-quality sampling. The amount of salt deposition or re-solution in the lake was estimated from changes in total dissolved salt load for the south and north parts of the lake during a given period of time. Precipitation of salt is indicated when the total load of dissolved salt decreased, and re-solution of salt is indicated when the total load of salt increased. All precipitation of dissolved salt since 1970 has occurred in the north part. The dissolved-solids concentration in the north part was at or near saturation during 1970–85. During 1986, lake volume increased to a volume sufficient to dissolve most if not all of the precipitated salt in the north part (fig. 4), and the total dissolved salt load peaked at about 4.9 billion tons; therefore, total salt load in the lake (dissolved plus precipitated) was about 4.9 billion tons during 1980–86. Previous investigators had estimated total salt load for the entire lake to be about 5.5 billion tons, by means of cores taken in the south and north parts by the UGS during the fall of 1970 and 1972.

Re-solution of salt in either the south (*LSD*) or north part (*LND*) is represented in tons by the equations revised from Waddell and Bolke (1973) as follows:

$$LSD=T[(483)(VS)-LS(I)](5.25 \times 10^{-3}) \quad (B4)$$

$$LND=T[(483)(VN)-LN(I)](5.25 \times 10^{-3}) \quad (B5)$$

where

LSD = redissolved salt in south part, in tons;
LND = redissolved salt in north part, in tons; and
T = time interval (here = 1.901 days), in days.

The constant 483 is an empirical constant associated with the limiting salt load in either part of the lake for a given volume, in tons per acre-foot. The constant is the product of the saturation concentration of sodium chloride (355 grams per liter) and a factor for converting from grams per liter to tons per acre-foot (355 x 1.36 = 483);

VS = volume of south part, in acre-feet;
VN = volume of north part, in acre-feet;
LS = dissolved salt load in south part at time step *I*, in tons;
LN = dissolved salt load in north part at time step *I*, in tons;
I = present time step; and
5.25×10^{-3} is an empirical constant for re-solution rate per day, in tons per day per total tons of salt deposited in north or south part.

Salt precipitation in the north part (*LNP*) was computed the same way in the causeway model as Waddell and Bolke (1973, p. 34) using the equation

$$LNP = LN - 483(VN) \quad (B6)$$

where

LNP = precipitated salt load in the north part, in tons.

Salt precipitation for the south part (*LSP*) was not calculated because a salt crust has not been observed since at least 1970.

The amount of salt precipitation that might occur in the future depends on inflow conditions and changes in the conveyance properties of the causeway. A decrease in inflow causes a decrease in lake volume, which can cause salt precipitation in the south and north parts. The distribution of salt precipitation between the south and north parts also is affected by the conveyance properties of the causeway. For a given volume, the smaller the openings in the causeway, the more likely precipitation of salt in the north part is to occur.

Stratification in Great Salt Lake

Before the causeway was built, the available data were not adequate to determine whether the lake was stratified. During and after completion of the causeway, data were collected (Madison, 1970; Waddell and Bolke, 1973; and Gwynn and Sturm, 1987) to define the stratification in the south and north parts of the lake.

Madison (1970, p. 12) observed that a lower layer of dense brine (*LSL*, fig. 7) occurred in the south part of the lake below 4,175 ft (fig. B1, August 1967 to June 1969 density profiles). Madison further noted that the volume of the lower layer of brine remained relatively constant and that this apparent stability is the result of equilibrium between the amount of dense brine moving from the north part of the lake south through the causeway and the amount of mixing that took place at the interface. Waddell and Bolke (1973) collected additional data in the south part during 1971–72 (fig. B1, May 1970 to May 1972 density profiles) that indicated this volume was about the same as during the study by Madison (1970).

Gwynn and Sturm (1987) observed changes in stratification during 1984 that indicated the interface in the south part became more diffused (fig. B1, July 1984 density profiles) as a result of record surface-water inflow during 1983–84. The interface in the south part became less diffused in 1986 (fig. B1, September 1986 density profiles) as a result of increased flow through the breach during August 1984 to September 1986. The transition from no stratification to gradual stratification in the north part (fig. B2, October 1980 to September 1986 density profiles) resulted from high inflow and the rapid rise of the water-surface altitude (Gwynn and Sturm, 1987). J.W. Gwynn (Utah Geological Survey, written commun., 1994) noted that the lake became essentially mixed with little or no stratification in either the south or north part by 1991.

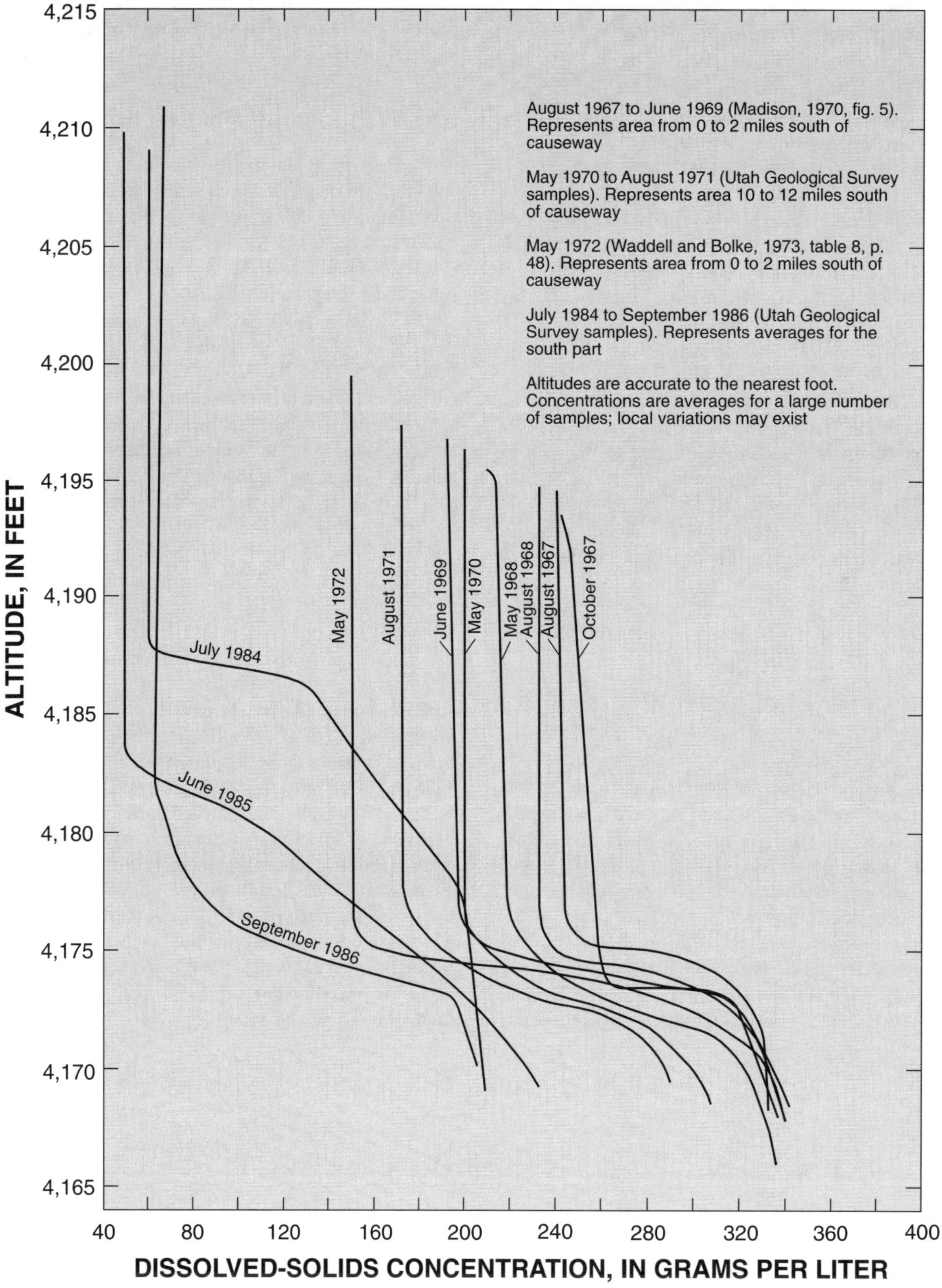

Figure B1. Approximate dissolved-solids concentration gradients for south part of Great Salt Lake, Utah, on selected dates, 1967–86.

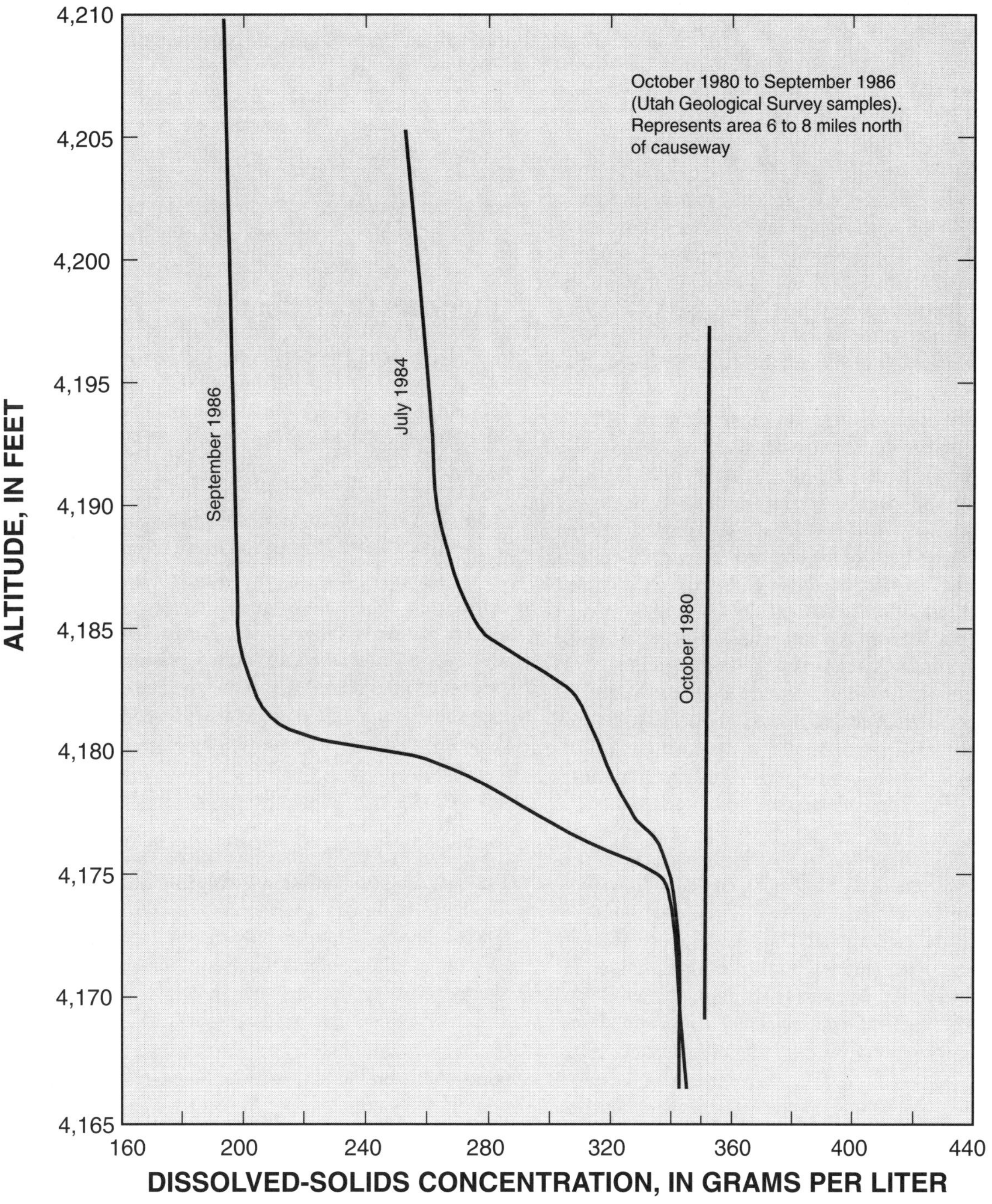

Figure B2. Approximate dissolved-solids concentration gradients for north part of Great Salt Lake, Utah, in October 1980, July 1984, and September 1986.

APPENDIX C. FLOW THROUGH THE CAUSEWAY FILL

In the early 1900s, fill material was used to form approaches to a railroad trestle that crossed Great Salt Lake. The approaches and trestle traversed the lake in an east-west direction from Promontory Point to Lakeside (fig. 1) where the lake is about 18 mi wide. The approaches have a combined length of about 6 mi and now abut either end of the causeway that was constructed during 1957–59 (fig. 2) to replace the trestle. The causeway was constructed across the remaining distance of 12.21 mi (Madison, 1970, p. 7) to support the railroad track above the surface of Great Salt Lake.

Construction of the causeway began by dredging a channel 25 to 40 ft deep and 150 to 500 ft wide. The channel was backfilled with sand and gravel, and quarry-run rock was used as fill to raise the top of the causeway above the surface of the lake. The causeway was completed with riprap that varied in size from 1-ton capstone 15 ft below the surface to 3-ton capstone at the top. The causeway is permeable and also was breached by two culverts, each 15 ft wide, that allowed brine to flow through open channels as well as through the fill material. Prior to construction of the causeway, circulation was unimpeded through the trestle. The causeway restricted circulation in the lake, but allowed brine to flow through the permeable fill material.

Two-way flow of brine occurs through the causeway fill (fig. 3) because of head and density differences between the south and north parts of the lake (ΔH and $\Delta \rho$). South-to-north flow (*QSF*) through the upper part of the fill is caused by a positive head difference (ΔH) at the causeway between the south and north parts. A positive density difference ($\Delta \rho$) between the north and south parts creates the potential for north-to-south flow (*QNF*) through the lower part of the fill.

During the Waddell and Bolke (1973) study of Great Salt Lake, a model developed by Pinder and Cooper (1970) was calibrated and used to simulate flow through the causeway fill for a wide range of boundary conditions. After determining *QSF* and *QNF* for a wide range of possible boundary conditions, multiple regression analysis involving *QSF* and *QNF* as functions of the boundary conditions were used to develop regression equations. These regression equations were then incorporated into the original causeway model. Regression equations for fill flow were used in the original causeway model because the computation time on most computers was too long to operate the Pinder and Cooper (1970) model interactively with the original causeway model.

For this study, a more efficient steady-state cross-sectional finite-difference solute-transport model (referred to here as the fill-flow model) for density-dependent flow (Sanford and Konikow, 1985) was calibrated for a larger range of boundary conditions, a larger grid size, and a larger amount of input data. The changed boundary conditions were larger dimensions of the fill, a larger head difference, higher water-surface altitudes, smaller absolute densities, and a different density profile as a function of depth for both parts of the lake.

The causeway fill was simulated using a finite-difference grid consisting of 45 cells in the vertical direction and 43 cells in the horizontal direction. Each cell represented possible flow in the vertical direction discretized in 1-ft intervals and in the horizontal direction discretized in 5-ft intervals (fig. C1).

The fill-flow model was calibrated to lake conditions prior to the breach of the causeway in August 1984. Most of the model simulations approximate conditions prior to the breach. After the breach, the density stratification of both parts of the lake changed. A few simulations were done with boundary conditions that occurred after the breach.

Boundary Conditions

The boundary conditions for the causeway affect the rate of movement of brine through the causeway fill. Waddell and Bolke (1973, table 4) determined that water moves through the center 12.21 mi of the causeway, which was built during 1957–59. The older part of the causeway that was built in the early 1900s abuts each end of the newer fill. The older fill is virtually impermeable relative to the more recent fill, and no water is assumed to move through the older fill.

Since 1970, sand, gravel, and rock have been added periodically to the causeway to increase height and to replace material removed by erosion from wave action. During 1982–84 when the lake surface rose almost 11 ft, a large amount of fill material was added to keep the causeway surface above water. Fill material also was added parallel to the causeway to compensate for settling of the causeway into the lake-bottom muds. From 1970 to 1987, the height of the fill increased by an average of about 7 ft, and width increased about 10 to 20 ft.

For this study, an average cross section of the fill in 1987 was determined using surveyed profiles (D. Warnock, Southern Pacific Transportation Co., written commun., 1987). Eight profiles at evenly spaced intervals of about 1.5 mi were used to compute the average cross section. The average cross section is shown in figure C1.

The upper boundary for flow of water through the fill was where the water surface meets the surface of the fill at atmospheric pressure. The lower boundary was simulated as no flow at an altitude of 4,170 ft. Dye studies done in 1971–72 determined that flow was insignificant below 4,170 ft (Waddell and Bolke, 1973, p. 22). On the south and north sides of the causeway fill, the lateral boundary conditions are (1) difference in water-surface altitude of the two parts, (2) total depth of water of the two parts, and (3) difference in density between the two parts. The different combinations of these three boundary conditions cause two-layer stratified movement of water in opposite directions. Many simulations were made with different combinations of the three boundary conditions to simulate fill flow for conditions that occurred during 1959–86 and for conditions that may occur in the future.

In the fill-flow model, constant-pressure boundaries were used on each side of the fill. Pressure for each cell on the boundary was computed from depth of water times the density of the column of overlying water. The constant-pressure boundaries also act as sources of a solute, in this case, a source of the density-controlling solute, total dissolved salt.

From 1965 until about 1991, the south part of the lake had a two-layer stratified density profile. The lake became essentially mixed with little or no stratification in either south or north part by 1991 (J.W. Gwynn, Utah Geological Survey, written commun., 1994).

The assumptions for all the pre-breach simulations are (1) The upper part has uniform density from the water surface to an altitude of 4,177.5 ft, (2) A 4-ft diffusion zone is between the upper and lower layers of different densities, and (3) The lower layer is from 4,173.5 ft to the bottom at 4,170 ft, and its density is uniform and equal to the density of the north part.

The north part of the lake has a uniform density profile except for the upper few feet. The upper few feet of the lake near the causeway has less dense water because the less dense water that flows from south to north spreads out on the surface of the north part. For all simulations, the upper 3 ft of the density profile was graded to a lesser density using the same gradient used for the 4-ft diffusion zone in the south part.

An example of the pre-breach density profiles for the south and north boundaries used in one of the simulations is shown in figures C2 and C3. Boundary conditions were a water-surface altitude of 4,198.0 ft in the north part, a head difference at the causeway of 1.5 ft, and a density difference of 0.124 g/mL. The boundary for the south part has a density of 1.086 g/mL from the water surface to an altitude of 4,177.5 ft (fig. 18). The 4-ft diffusion zone is represented at 1-ft intervals from 4,177.5 to 4,173.5 ft with densities of 1.086, 1.115, 1.149, 1.176, and 1.210 g/mL. The lower layer has the same density as the north part, 1.210 g/mL. The boundary for the north part has a density of 1.210 g/mL from 4,194.5 ft to the lake bottom, about 4,170 ft (fig. 19). The 3-ft diffusion zone is represented at 1-ft intervals from 4,197.5 to 4,194.5 ft with densities of 1.115, 1.149, 1.176, and 1.210 g/mL. The density difference of 0.124 g/mL between the two parts is computed by subtracting the density of the upper layer of the south part (1.086 g/mL) from the density of the lower layer of the north part (1.210 g/mL).

There were 125 simulations done with different head and density differences and different water-surface altitudes for the north part. All simulations had the same 4-ft diffusion zone in the south part with the same altitude for the dif-

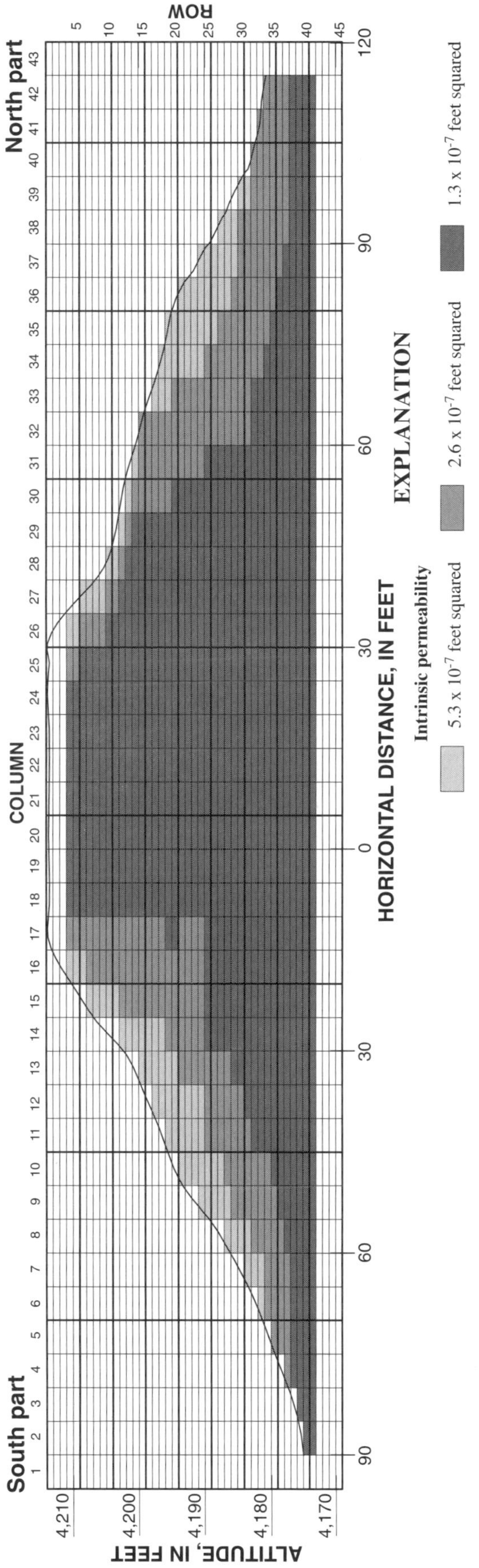

Figure C1. Fill-flow model grid of the cross section of causeway fill across Great Salt Lake, Utah, showing intrinsic permeability values of cells.

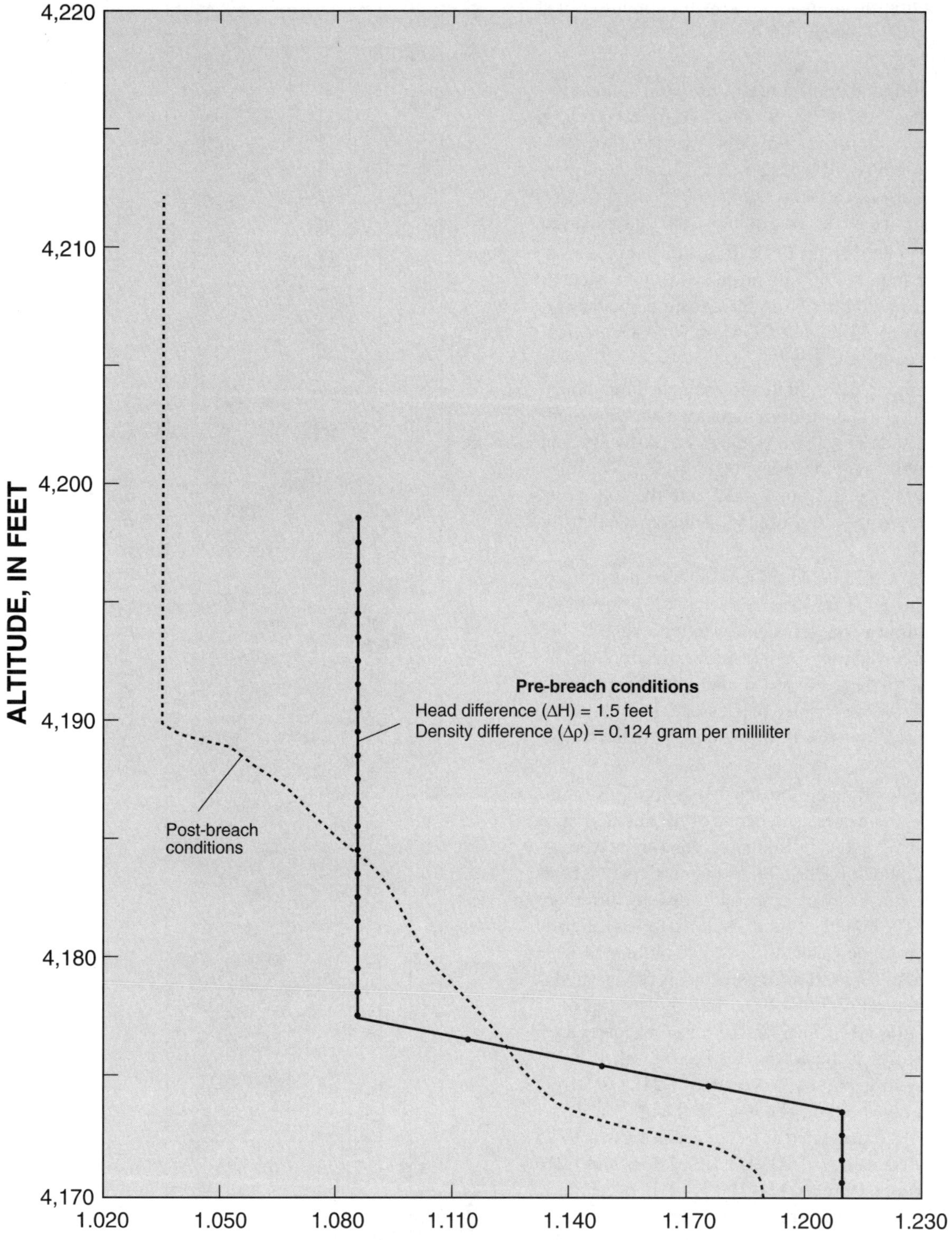

Figure C2. Density profile used in fill-flow model for south part of Great Salt Lake, Utah. Post-breach conditions from Gwynn and Sturm (1987, fig. 12).

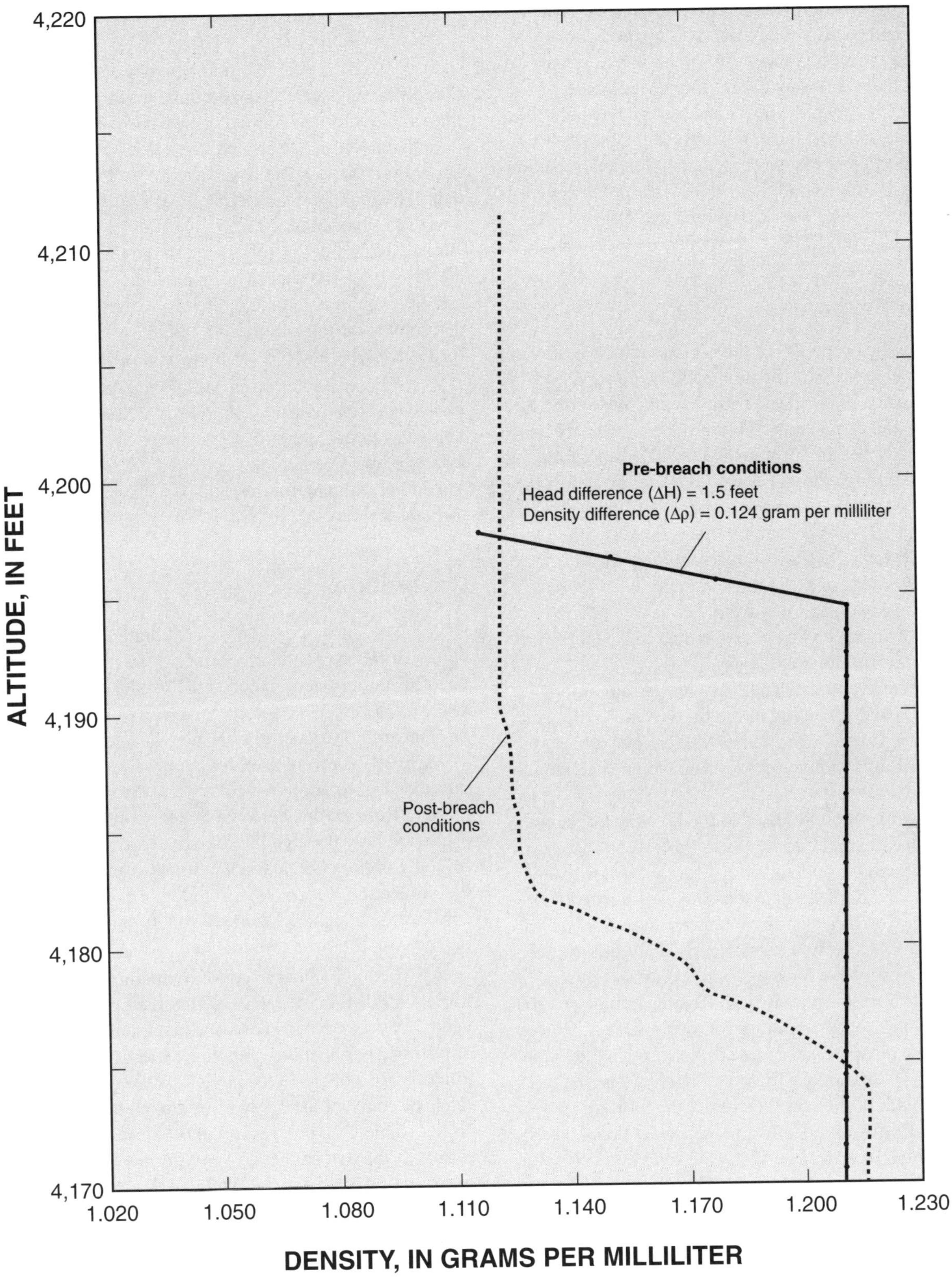

Figure C3. Density profile used in fill-flow model for north part of Great Salt Lake, Utah. Post-breach conditions from Gwynn and Sturm (1987, fig. 13).

fusion zone, and the same 3-ft diffusion zone at the upper part of the north boundary.

The density stratification for the south and north part of the lake changed after the breach was opened. Post-breach density profiles for the south and north part are shown in figures C2 and C3, respectively. In the south part, the sharp interface between the upper and lower layers of different density changed to a much more gradual zone of diffusion. In the north part, the uniform density to depth relation changed to a stratified condition. These two profiles were fairly constant during 1986–87 (Gwynn and Sturm, 1987, figs. 11–14a; Gwynn, 1988, Appendix A).

Hydraulic Properties

Hydraulic properties of the fill that influence flow of water and solutes are intrinsic permeability, dispersivity, effective porosity, molecular diffusivity, and anisotropy for permeability and dispersivity. Hydraulic conductivity and effective porosity of the fill were estimated in a previous study of flow through the causeway fill (Waddell and Bolke, 1973, table 3). Those estimates were made from dye-injection studies in 10 wells drilled into the causeway fill. Hydraulic conductivity was estimated to range from 0.08 to 2.10 ft/s. That range of hydraulic conductivity converts to a range of intrinsic permeability from 2.68 x 10^{-8} to 7.04 x 10^{-7} ft ^{2}. Effective porosity was estimated to be a uniform value of 0.3 for the entire causeway.

Dispersivity, molecular diffusivity, and anisotropy for permeability and dispersivity of the fill were not estimated from those dye studies. The values of these properties as used in the fill-flow model were determined during calibration and sensitivity analysis.

Hydraulic properties used in the fill-flow model are intrinsic permeability, effective porosity, dispersivity, molecular diffusivity, and anisotropy for permeability and dispersivity. The fill was simulated in a two-dimensional cross section. The areal distribution of permeability was varied within the cross section; however, the 12.21 mi of causeway fill was assumed to have the same cross-sectional permeability along its entire length. The magnitude and distribution of intrinsic permeability for the cross section used in the fill-flow model was the same as that used in the Pinder and Cooper (1970) model. The only difference is that the height and width of the cross section of the fill is slightly larger in the fill-flow model than in the Pinder and Cooper (1970) model. The new material on the sides and above the old core of the fill was assigned the permeability values used on the outside of the fill in the Pinder and Cooper (1970) model. The values of hydraulic conductivity in the Pinder and Cooper (1970) model were 0.39, 0.79, and 1.57 ft/s. These values were converted to intrinsic permeability values of 1.3 x 10^{-7}, 2.6 x 10^{-7}, and 5.3 x 10^{-7} ft ^{2} in the fill-flow model: 5.3 x 10^{-7} ft^2 forming the outer layer of the fill about 10 ft thick, 2.6 x 10^{-7} ft^2 forming the middle layer about 15 ft thick, and 1.3 x 10^{-7} ft^2 forming the inner core. The inner core of smaller permeability ranges through widths of 35 ft at an altitude of 4,210 ft, 60 ft at an altitude of 4,200 ft, and 130 ft at an altitude of 4,180 ft.

Values of effective porosity, dispersivity, and molecular diffusivity were assumed to be uniform for all the fill material. Estimates of these properties were made for the initial simulations. If needed, the values were modified during calibration to achieve agreement with measured flow data. The initial value of effective porosity was 0.3, which was the value estimated from dye studies (Waddell and Bolke, 1973) and the value used in the Pinder and Cooper (1970) model. Estimates of dispersivity and molecular diffusivity were not available; therefore, these properties were determined during calibration. Initial estimates were 10.0 ft for dispersivity and 0 ft for molecular diffusivity.

Anisotropy for permeability and dispersivity also were determined during calibration. These properties are treated as ratios in the fill-flow model. The initial estimate for the ratio of vertical to horizontal permeability was 1.0. The initial estimate for the ratio of transverse to longitudinal dispersivity was 0.50.

Calibration

The fill-flow model was calibrated by changing the values of the hydraulic properties of the fill until the simulated fill-flow system matched the results of the dye studies made in 1971–72. In addition, the computed fill flows from the Pinder and Cooper (1970) model (Waddell and Bolke, 1973, p. 20–31) were used for comparison. The dye studies provided information on velocity and density profiles, width of the diffusion zone between large and small densities, and estimated flow through the fill. Two dye studies were done for two different boundary conditions. Computed flow from the Pinder and Cooper (1970) model was used for comparison because that model worked well in simulations of the salt balance for 1969–72 (Waddell and Bolke, 1973, fig. 3).

The final values used for hydraulic properties of the fill in the fill-flow model are within the range of estimated values from the dye studies. The magnitude and distribution of intrinsic permeability were the same as those used in the Pinder and Cooper (1970) model. During calibration, a range of constant factors was multiplied by intrinsic permeability (table C1), but the previous values best matched the results of the dye studies. The uniform value for effective porosity of 0.3 used in the Pinder and Cooper (1970) model also was the final value used in the fill-flow model.

Dispersivity was determined by starting with a value of 10.0 ft and incrementally decreasing the value until the fill-flow model-computed vertical distributions of density matched the profiles measured in the dye studies. The value of longitudinal dispersivity used in the calibrated fill-flow model was 0.05 ft (table C2). Molecular diffusivity esti-

Table C1. Sensitivity of selected diffusion-zone widths in the cross section of the causeway fill for changes in intrinsic permeability, Great Salt Lake, Utah

[ΔH, head difference at the causeway; Δρ, density difference; EN, water-surface altitude of the north part; QSF, south-to-north flow through fill; QNF, north-to-south flow through fill; ft, feet; g/mL, grams per milliliter; ft ^{3}/s, cubic feet per second; NA, not applicable]

Boundary conditions				South-to-north flow		North-to-south flow		Width of diffusion zone (ft)		
								Column number		
ΔH (ft)	Δρ (g/mL)	EN (ft)	Permeability factor[1]	QSF (ft^3/s)	Calibrated	QNF (ft^3/s)	Calibrated	15	20	25
1.5	0.124	4,198.0	1.0 calibrated	4,500	NA	945	NA	6	6	7
			[2].5	2,250	.50	472	.50	6	6	7
			[3].1	443	.098	93	.098	6	6	7
			2.0	9,000	2.00	1,950	2.06	6	6	8
.5	.124	4,204.0	1.0 calibrated	1,120	NA	5,350	NA	8	9	10
			[2].5	560	.50	2,670	.50	8	8	10
			2.0	2,220	1.98	10,600	1.98	8	8	10
.5	.06	4,199.0	1.0 calibrated	860	NA	920	NA	6	7	6
			[2].5	428	.50	459	.50	6	7	6
			2.0	1,700	1.98	1,820	1.98	6	6	6

[1]Multiplied by the calibrated intrinsic permeability for the three zones with the following values shown in figure 17: 5.3 x 10^{-7} feet squared, 2.6 x 10^{-7} feet squared, 1.3 x 10^{-7} feet squared.

[2]Simulation time to reach equilibrium (6 hours) was two times that of calibrated fill-flow model (3 hours).

[3]Simulation time to reach equilibrium (12 hours) was four times that of calibrated fill-flow model.

mated as zero provided adequate results and was used in the final calibrated fill-flow model.

Anisotropy of permeability used in the fill-flow model was a ratio of vertical to horizontal permeability and the calibrated value was 1.0. Anisotropy of dispersivity used in the fill-flow model was a ratio of transverse to longitudinal dispersivity and the calibrated value was 0.50.

Profiles of velocity and density computed from the calibrated fill-flow model agree with measured profiles obtained from the dye studies from seven wells (figs. C4 and C5). The measured profiles are variable, but the computed profiles are similar to the average profile of the seven wells. Some of the profiles vary near the north edge of the fill; taking the measurement a few feet south or north of the fill makes a large difference in the profile.

Using the results of the dye studies, flow through the fill was estimated for two different boundary conditions (Waddell and Bolke, 1973, table 4 and p. 22–28). In late August and early September 1971, density difference was 0.110 g/mL, head difference at the causeway was 1.0 ft, and water-surface altitude of the north part was 4,196 ft. Estimated south-to-north flow was 1,600 ft ^{3}/s. No estimate was made for north-to-south flow. Using the same boundary conditions, the calibrated fill-flow model computed south-to-north flow to be 1,540 ft^3/s. In late May and early June 1972, density difference was 0.124 g/mL, head difference at the

Table C2. Sensitivity of selected diffusion-zone widths in the cross section of the causeway fill for changes in longitudinal dispersivity, Great Salt Lake, Utah

[Boundary conditions: Head difference at causeway (ΔH) = 1.5 feet; Density difference (Δρ) = 0.124 gram per liter; Density of brine in south part (ρs) = 1.086 grams per liter; Density of brine in north part (ρn) = 1.210 grams per liter; Water-surface altitude of north part (EN) = 4,198.0 feet; Width of diffusion zone determined between cells with density less than or equal to 1.088 grams per milliliter and density greater than or equal to 1.209 grams per milliliter]

Longitudinal dispersivity, in feet	Width of diffusion zone, in feet				
	Column number				
	10	15	20	25	29
0.01	4	4	4	5	5
.05	6	6	6	7	7
.10	7	7	8	9	9
.50	12	13	14	16	(1)
1.00	(1)	(1)	20	(1)	(1)

[1]Diffusion zone extends to upper or lower boundary.

Boundary conditions			
Density difference (Δρ, in grams per milliliter)	Head difference (ΔH, in feet)	Water-surface altitude in north part (EN, in feet)	Velocity profile
0.124	1.50	4,198.0	Profile from fill-flow model
.122	1.62	4,197.8	Profile from observation well 1
.125	1.59	4,197.8	Profile from observation well 3
.121	1.69	4,197.8	Profile from observation well 4
.108	1.62	4,197.8	Profile from observation well 5
.122	1.60	4,197.8	Profile from observation well 8
.122	1.63	4,197.8	Profile from observation well 9
.122	1.53	4,197.8	Profile from observation well 10

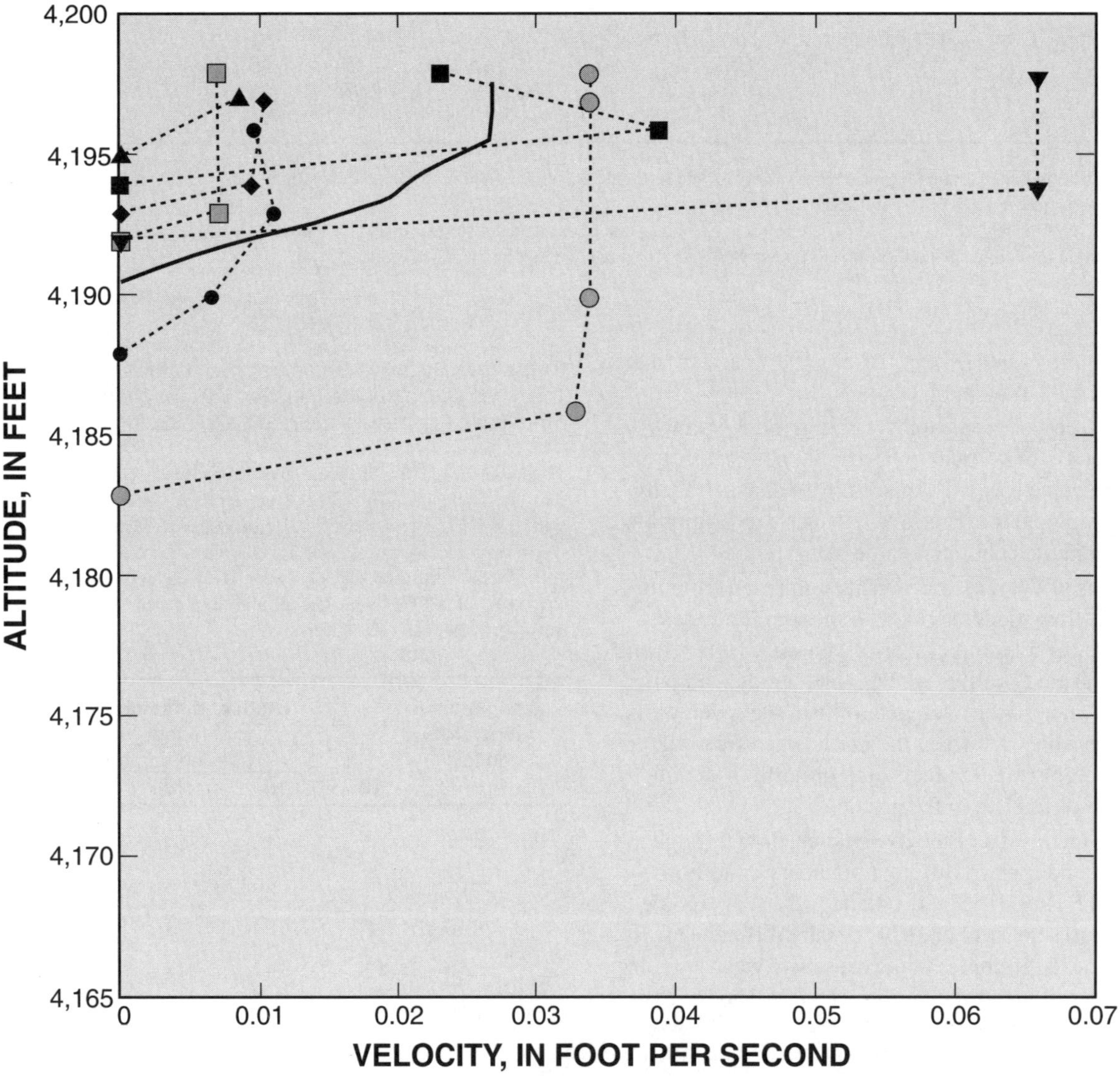

Figure C4. Velocity profiles from the fill-flow model and from wells in the causeway across Great Salt Lake, Utah. Velocity profiles from observation wells are from Waddell and Bolke (1973, figs. 1 and 9).

Boundary conditions			
Density difference (Δρ, in grams per milliliter)	Head difference (ΔH, in feet)	Water-surface altitude in north part (EN, in feet)	Density profile
0.124	1.50	4,198.0	Profile from fill-flow model
.122	1.62	4,197.8	Profile from observation well 1
.125	1.59	4,197.8	Profile from observation well 3
.121	1.69	4,197.8	Profile from observation well 4
.108	1.62	4,197.8	Profile from observation well 5
.122	1.60	4,197.8	Profile from observation well 8
.122	1.63	4,197.8	Profile from observation well 9
.122	1.53	4,197.8	Profile from observation well 10

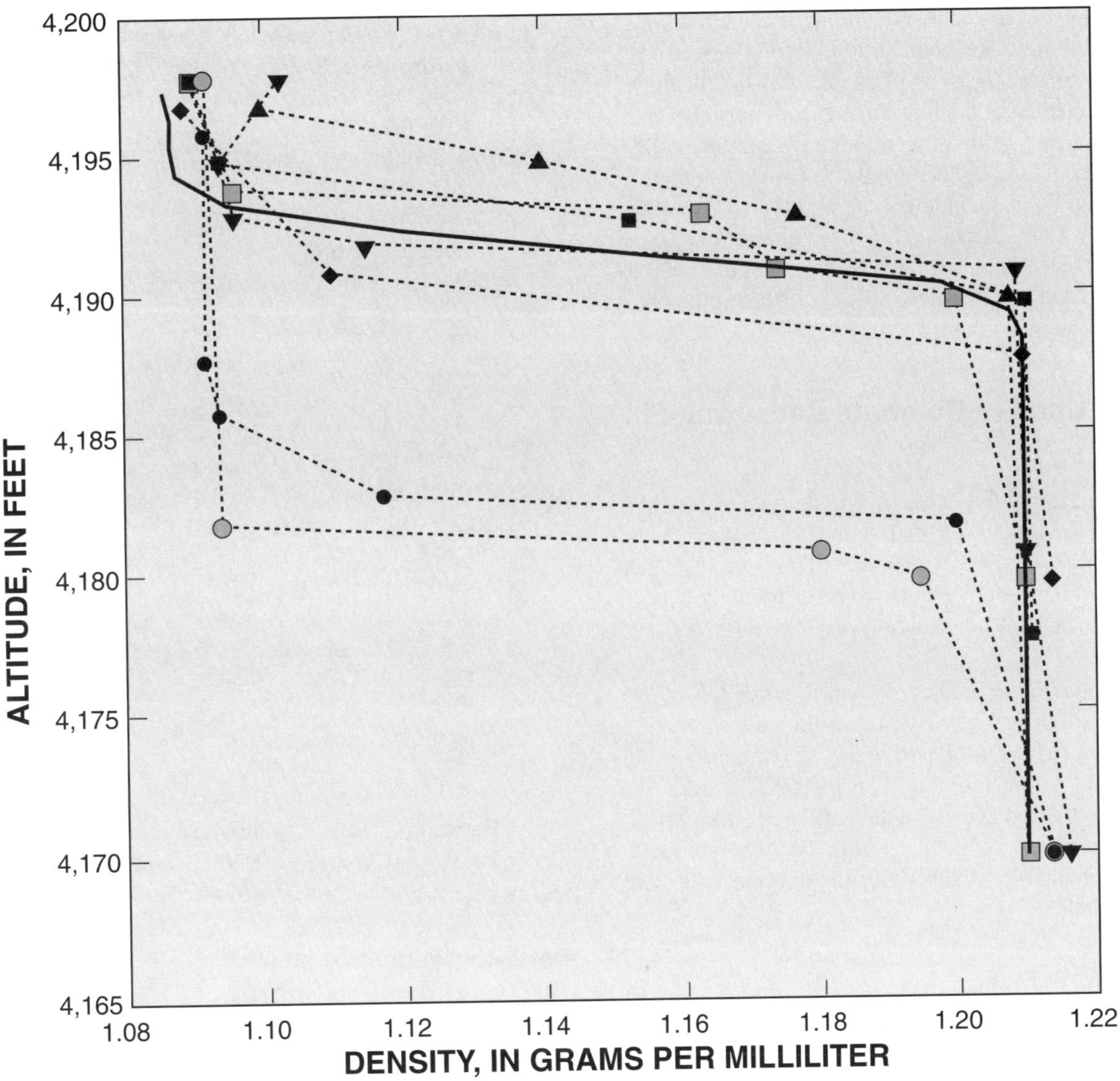

Figure C5. Density profiles from the fill-flow model and from wells in the causeway across Great Salt Lake, Utah. Density profiles from observation wells are from Waddell and Bolke (1973, figs. 1 and 9).

causeway was 1.5 ft, and water-surface altitude of the north part was 4,198.0 ft. Estimated south-to-north flow was 4,500 ft^3/s and estimated north-to-south flow was 1,400 ft^3/s. Computed south-to-north flow was 4,500 ft^3/s, and computed north-to-south flow was 945 ft^3/s.

A cross section showing contours of computed brine velocity through the fill for the boundary conditions of density difference ($\Delta\rho$) of 0.124 g/mL, head difference (ΔH) at the causeway of 1.5 ft, and water-surface altitude in the north part (*EN*) of 4,198.0 ft is shown in figure C6. The two-way flow is clearly shown with south-to-north flow moving through the upper part of the fill and north-to-south flow moving through the lower part of the fill.

Selected Causeway-Fill Flows

The fill-flow model was used to compute selected fill flows for a range of boundary conditions, and the relations between the fill flows and boundary conditions were used in the causeway model. Flow was computed by the fill-flow model as an average of the flow through columns 16–20 of the finite-difference grid. Flow through the causeway fill (tables C3–C5) was computed for the following range of boundary conditions, including those observed on the lake from the early 1960s through 1986: head difference (ΔH) at the causeway of 0.10 to 3.9 ft, density difference ($\Delta\rho$) of 0.02 to 0.15 g/mL, and water-surface altitude of the north part (*EN*) of 4,191 to 4,211 ft.

Methods Used to Compute Causeway-Fill Flow

Graphical methods were used to determine possible functional relations between fill flow and principal variables. These methods were used to compute fill flow because computation time on most computers would be too long to operate the fill-flow model interactively with the causeway model.

The following estimates of south-to-north flow (*QSF*) and north-to-south flow (*QNF*) through the fill were made with the density of the north part of the lake primarily equal to 1.210 g/mL. Estimated flow, *QSF* and *QNF*, is valid for the range of observed density in the north part (from 1.120 g/mL to saturation, 1.223 g/mL) within the inaccuracies of the values of hydraulic conductivity and/or porosity used in the fill-flow model.

South-to-North Flow

A suitable functional relation for south-to-north flow through the fill (*QSF*) in terms of head difference (ΔH) at the causeway, density of the south part (ρs), and average thickness of the upper-brine layer (*YSF*), which flows from south-to-north, could not be determined. Therefore, *QSF* was computed by linearly interpolating between discrete boundary conditions (ΔH, $\Delta\rho$, and *EN*) for flow computed by the fill-flow model when the density of the north part (ρn) was 1.210 g/mL (table C6).

The following calculation illustrates how south-to-north flow (*QSF*), which varies with ΔH, *EN*, and $\Delta\rho$, was interpolated from data in table C6 for a specific set of boundary conditions (table C7). The boundary conditions for the calculation are $\Delta H = 0.30$ ft, $\Delta\rho = 0.050$ g/mL, and $EN =$ 4,200 ft.

The interpolation necessary for $QSF_{\Delta H}$ is shown in equation C1.

$$QSF_{0.30}=(QSF_{0.50}-QSF_{0.25})\cdot(0.30-0.25)/(0.50-0.25) + QSF_{0.25} \qquad (C1)$$

Equation 20 is used to interpolate the following flows at $\Delta H = 0.30$ ft and their respective *EN* and $\Delta\rho$.

At $EN = 4{,}199$ ft and $\Delta\rho = 0.02$ g/mL: $753 = (1{,}653-528)\cdot 0.2 + 528$.

At $EN = 4{,}204$ ft and $\Delta\rho = 0.02$ g/mL: $1{,}135 = (2{,}273-850)\cdot 0.2 + 850$.

At $EN = 4{,}199$ ft and $\Delta\rho = 0.06$ g/mL: $433 = (856-327)\cdot 0.2 + 327$.

At $EN = 4{,}204$ ft and $\Delta\rho = 0.06$ g/mL: $711 = (1{,}295-565)\cdot 0.2 + 565$.

The interpolation necessary for QSF_{EN} is shown in equation C2.

$$(QSF_{4,204} - QSF_{4,199})\cdot(4{,}200-4{,}199)/(4{,}204-4{,}199) + QSF_{4,199} = QSF_{4,200} \qquad (C2)$$

Equation C2 is used to interpolate the following flows at $\Delta H = 0.30$ ft, $EN = 4{,}200$ ft, and the respective $\Delta\rho$.

For $\Delta\rho = 0.02$ g/mL: $QSF_{4,200} = (1{,}135-753)\cdot 0.2 + 753 = 829$.

For $\Delta\rho = 0.06$ g/mL: $QSF_{4,200} = (711-433)\cdot 0.2 + 433 = 488$.

The interpolation necessary for $QSF_{\Delta\rho}$ is shown in equation C3.

$$QSF_{0.05}=(QSF_{0.06}-QSF_{0.02})\cdot(0.05-0.02)/(0.06-0.02) + QSF_{0.02} \qquad (C3)$$

Equation 22 is used to interpolate final flow, *QSF*.

For $\Delta H = 0.30$ ft, $EN = 4{,}200$ ft, and $\Delta\rho = 0.05$ g/mL: $QSF = (488-829)\cdot 0.75 + 829 = 574\ ft^3/s$.

North-to-South Flow

An empirical equation to compute north-to-south flow (*QNF*) was developed using the Ghyben-Herzberg principle (Badon-Ghyben, 1888; and Herzberg, 1901). Based on the Ghyben-Herzberg principle, equation C4 is used to compute the thickness of the lower brine layer (*YNF*), which flows from north to south.

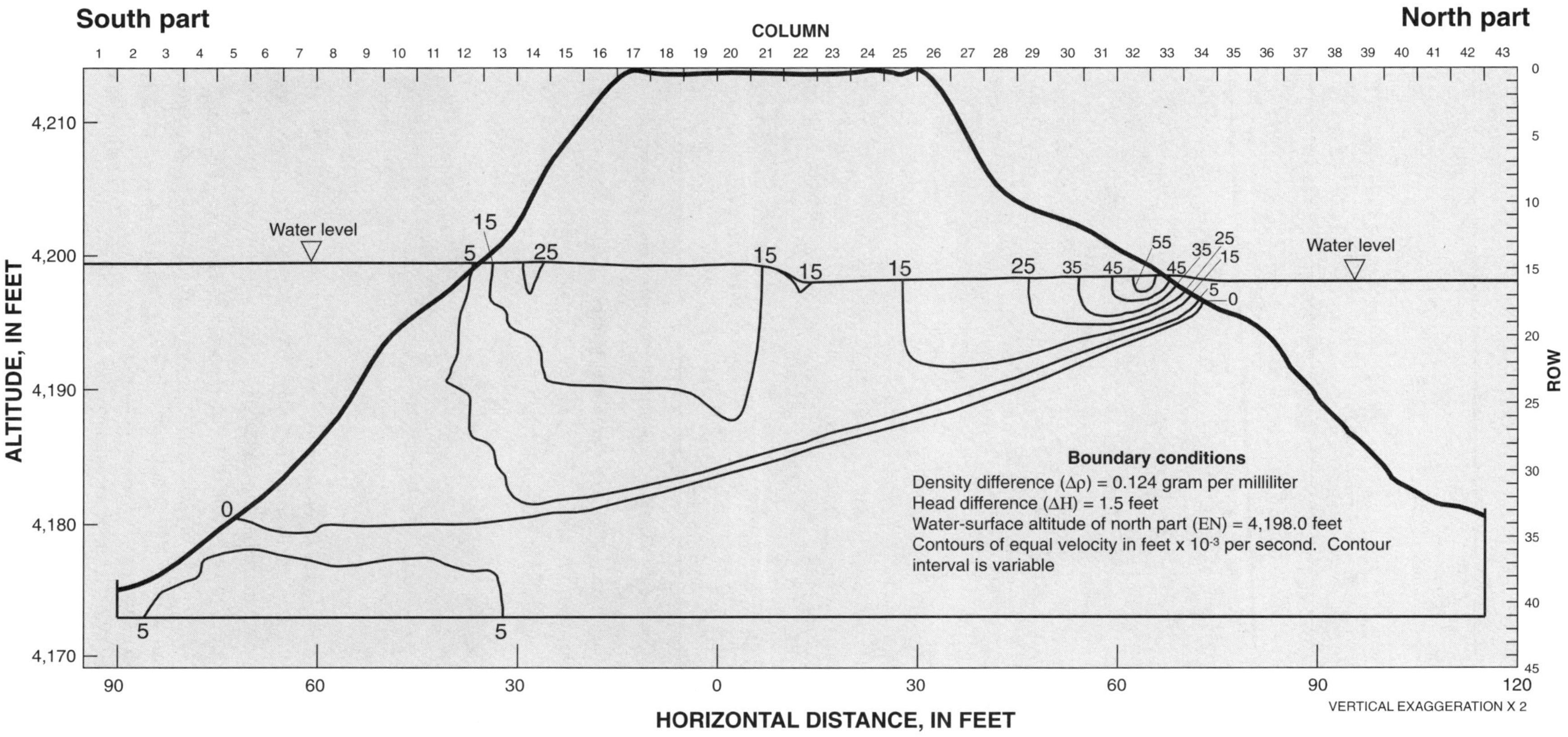

Figure C6. Cross section of brine velocity through causeway fill for specific boundary conditions, Great Salt Lake, Utah.

Table C3. South-to-north flow (QSF) through the causeway fill as computed by the fill-flow model, Great Salt Lake, Utah

[*EN*, water-surface altitude of north part; flow in cubic feet per second; density of north part is 1.210 grams per milliliter except where noted; —, no data]

Density difference ($\Delta\rho$), in grams per milliliter	EN, in feet	Head difference at the causeway (ΔH), in feet										
		0.10	0.25	0.50	0.75	1.00	1.50	1.75	2.00	2.50	3.00	3.50
0.02	4,191	59	—	1,086	—	2,590	4,677	—	6,063	—	—	—
.02	4,194	87	—	1,264	—	3,129	5,487	—	7,449	—	—	—
.02	4,199	112	528	1,653	—	4,390	7,406	8,765	10,367	13,232	18,010	
.02	4,204	134	—	2,273	—	6,398	11,290	—	15,920	—	—	—
.06	4,191	3	—	561	—	1,813	4,292	—	5,908	—	—	—
.06	4,194	13	—	653	—	2,002	4,817	—	7,046	—	—	—
.06	4,199	92	327	856	1,577	2,465	6,046	7,470	9,203	12,600	17,930	
.06	4,204	168	—	1,295	—	3,599	8,419	—	13,490	—	—	—
.10	4,191	9	—	447	—	1,284	3,684	—	5,320	—	—	—
.10	4,194	71	—	568	—	1,472	4,093	—	6,059	—	—	—
.10	4,199	142	303	729	—	1,976	4,901	5,908	7,327	10,850	16,760	—
		—		—		[1]2,128	—			[1]11,550		
.10	4,204	172	—	1,102	—	2,925	7,027	—	10,830	16,050	22,360	27,570
						[1]3,181				[1]17,110		
.124	4,191	9	—	418	—	1,195	3,403	—	4,914	—	—	—
.124	4,194	60	—	505	—	1,374	3,759	—	5,466	—	—	—
.124	4,199	141	313	737	—	1,873	4,683	5,524	6,739	9,772	15,580	
						[2]2,009				[2]10,500		
.124	4,204	184	—	1,125	—	2,799	6,782	—	10,210	15,000	20,590	25,520
						[2]3,041				[2]16,110		

[1]Flow computed with lower absolute density. Density of brine in south part (ρs) = 1.063 grams per milliliter; density of brine in north part (ρn) = 1.163 grams per milliliter.

[2]Flow computed with lower absolute density. Density of brine in south part (ρs) = 1.039 grams per milliliter; density of brine in north part (ρn) = 1.163 grams per milliliter.

$$YNF = EN - 4{,}175 - \Delta H \bullet \rho s/\Delta\rho \qquad (C4)$$

where

YNF = average thickness of the lower layer of brine flowing through fill from north to south, in feet;

EN = water-surface altitude of north part, in feet; and

4,175 = water-surface altitude of lower boundary of fill flow, in feet.

Plotting *QNF* as a function of $(YNF)^2$ resulted in a family of straight lines, each with a slope proportional to $\Delta\rho$. Equation C5 was developed for *QNF* relating the boundary conditions $\Delta\rho$ and *YNF*.

$$QNF = a(\Delta\rho)(YNF)^2 + b \qquad (C5)$$

where

QNF = north-to-south flow through the fill,

a = 73.401 for *QNF* less than or equal to 4,300 ft^3/s or 84.401 for *QNF* greater than 4,300 ft^3/s, and

b = 0.0 for *QNF* less than or equal to 4,300 ft^3/s or -516.54 for *QNF* greater than 4,300 ft^3/s.

The slope between *QNF* and $(YNF)^2$ is proportional to $\Delta\rho$. This empirical equation is advantageous because it has some physical basis and is a low-order polynomial. The model-computed north-to-south flow (tables C4–C5) and regression-computed north-to-south flow are shown in figure C7.

Table C4. North-to-south flow (QNF) through the causeway fill as computed by the fill-flow model, Great Salt Lake, Utah

[g/mL, grams per milliliter; EN, water-surface altitude of north part; flow in cubic feet per second; density of north part is 1.210 grams per milliliter except where noted; —, no data]

Density difference ($\Delta\rho$), in g/mL	EN, in feet	Head difference at the causeway (ΔH), in feet										
		0.10	0.25	0.50	0.75	1.00	1.50	1.75	2.00	2.50	3.00	3.50
0.02	4,191	154	—	0	—	0	0	—	0	—	—	—
.02	4,194	262	—	0	—	0	0	—	0	—	—	—
.02	4,199	494	141	0	—	0	0	—	0	0	0	—
.02	4,204	830	—	0	—	0	0	—	0	—	—	—
.06	4,191	831		195	—	0	0	—	0	—	—	—
.06	4,194	1,202		409	—	0	0	—	0	—	—	—
.06	4,199	2,110	1,505	918	441	94	0	0	0	0	0	—
.06	4,204	3,434		1,691	—	495	0	—	0	—	—	—
.10	4,191	1,532		742	—	188	0	—	0	—	—	—
.10	4,194	2,212		1,248	—	474	6	—	0	—	—	—
.10	4,199	3,720	2,960	2,280	—	1,202	450	170	0	0	0	—
						[1]1,465				[1]0		
.10	4,204	6,009		3,916	—	2,403	1,259	—	466	44	0	0
						[1]2,806				[1]106		
.124	4,191	1,977	—	1,117	—	498	61	—	0	—	—	—
.124	4,194	2,841	—	1,748	—	946	360	—	0	—	—	—
.124	4,199	4,719	3,865	3,182	—	2,042	1,134	777	433	29	0	
						[2]2,383				[2]134	—	—
.124	4,204	7,584	—	5,349	—	3,720	2,420	—	1,384	687	147	0
						[2]4,271				[2]901		

[1]Flow computed with lower absolute density. Density of brine in south part (ρs) = 1.063 grams per milliliter; density of brine in north part (ρn) = 1.163 grams per milliliter.

[2]Flow computed with lower absolute density. Density of brine in south part (ρs) = 1.039 grams per milliliter; density of brine in north part (ρn) = 1.163 grams per milliliter.

Table C5. Post-breach flow through the causeway fill as computed by the fill-flow model, Great Salt Lake, Utah

[g/mL, grams per milliliter; ft^3/s, cubic feet per second]

Density difference[1] (Δρ), in g/mL	Density of upper layer		Head difference (ΔH), in feet	Water-surface altitude of north part, (EN), in feet	Flow	
	South part, in g/mL	North part, in g/mL			South to north (QSF), in ft^3/s	North to south (QNF), in ft^3/s
0.085	1.035	1.120	0.5	4,206	1,307	4,107
.085	1.035	1.120	1.0	4,206	3,489	1,869
.085	1.035	1.120	.5	4,211	1,564	6,953
.085	1.035	1.120	1.0	4,211	4,161	3,863
.055	1.045	1.100	.5	4,206	1,508	1,942
.055	1.045	1.100	1.0	4,206	4,467	383
.055	1.045	1.100	.5	4,211	1,807	3,512
.055	1.045	1.100	1.0	4,211	5,234	1,039
.025	1.055	1.080	.5	4,206	2,387	322
.025	1.055	1.080	1.0	4,206	6,873	47
.025	1.055	1.080	.5	4,211	2,737	500
.025	1.055	1.080	1.0	4,211	8,547	94

[1] Three density differences were used in the post-breach flow computations. Density differences in the post-breach flow computations were computed as the difference between the upper layers of uniform density in the south and north parts. The lower layers of greater density in both parts were kept constant in all post-breach flow computations.

Table C6. Interpolation matrix for south-to-north flow through the causeway fill (QSF), Great Salt Lake, Utah

Density difference ($\Delta\rho$), in grams per milliliter	Water-surface altitude of north part (EN), in feet	Flow in cubic feet per second for indicated head difference (ΔH), in feet															
		0.00	0.10	0.25	0.50	0.75	1.00	1.25	1.50	1.75	2.00	2.25	2.50	2.75	3.00	3.25	3.50
0.020	4,191	0.00	59	465	1,086	1,770	2,590	3,700	4,677	5,430	6,063	6,900	8,150	10,190	12,280	13,810	15,060
.020	4,194	.00	87	500	1,264	2,135	3,129	4,330	5,487	6,570	7,449	8,480	9,900	12,100	14,260	15,900	17,210
.020	4,199	.00	112	528	1,653	2,990	4,390	5,970	7,406	8,765	10,367	11,700	13,232	15,750	18,010	19,810	21,340
.020	4,204	.00	134	850	2,273	4,170	6,398	8,870	11,290	13,630	15,920	18,290	20,770	23,140	25,600	27,330	29,220
.020	4,211	.00	135	1,149	2,576	4,885	8,052	11,070	14,457	17,804	20,536	22,425	25,185	27,749	30,354	32,039	33,861
.060	4,191	.00	3	200	561	1,025	1,813	3,050	4,292	5,225	5,908	6,710	8,120	10,050	12,250	13,760	15,030
.060	4,194	.00	13	230	653	1,150	2,002	3,330	4,817	6,050	7,046	8,140	9,750	11,860	14,170	15,810	17,170
.060	4,199	.00	92	327	856	1,577	2,465	4,190	6,046	7,470	9,203	10,750	12,600	15,100	17,930	19,670	21,010
.060	4,204	.00	168	565	1,295	2,200	3,599	5,775	8,419	11,000	13,490	16,190	19,130	21,670	24,330	26,330	28,450
.060	4,211	.00	245	779	1,495	2,449	4,317	6,352	9,355	12,926	15,780	18,003	21,039	23,569	26,165	27,996	29,902
.100	4,191	.00	9	146	447	700	1,284	2,430	3,684	4,600	5,320	6,290	7,870	10,000	12,160	13,620	15,000
.100	4,194	.00	71	248	568	820	1,472	2,700	4,093	5,200	6,059	7,140	8,960	11,430	13,840	15,570	17,070
.100	4,199	.00	142	303	729	1,240	1,976	3,095	4,901	5,908	7,327	8,880	10,850	13,700	16,760	18,860	20,680
.100	4,204	.00	172	520	1,102	1,780	2,925	4,850	7,027	9,000	10,830	13,300	16,050	19,200	22,360	25,000	27,570
.100	4,211	.00	215	645	1,353	2,164	3,519	5,774	8,278	10,490	12,487	15,169	18,104	21,418	24,663	27,263	29,720
.124	4,191	.00	9	160	418	680	1,195	2,240	3,403	4,260	4,914	5,710	7,120	9,240	11,350	12,860	14,080
.124	4,194	.00	60	200	505	780	1,374	2,500	3,759	4,720	5,466	6,380	7,920	10,290	12,630	14,290	15,660
.124	4,199	.00	141	313	737	1,120	1,873	3,280	4,683	5,524	6,739	8,000	9,772	12,500	15,580	17,620	19,320
.124	4,204	.00	184	560	1,125	1,720	2,799	4,680	6,782	8,600	10,210	12,390	15,000	17,800	20,590	23,050	25,520
.124	4,211	.00	221	670	1,334	2,023	3,264	5,410	7,772	9,770	11,496	13,827	16,590	19,509	22,361	24,802	27,204

Table C7. Summary of intermediate and final interpolated south-to-north flow (QSF) for example boundary conditions

[In this example $\Delta H = 0.30$ foot, $\Delta\rho = 0.050$ gram per milliliter, and $EN = 4{,}200$ feet. Values in brackets [] are interpolated values or boundary conditions; all other values are calculated values listed in table C6]

Density difference ($\Delta\rho$), in grams per milliliter	Water-surface altitude of north part (EN), in feet	Flow in cubic feet per second for indicated head difference (ΔH), in feet		
		0.25	[0.30]	0.50
0.020	4,199	528	753	1,653
.020	[4,200]		[829]	
.020	4,204	850	1,135	2,273
[.050]	[4,200]		[574]	
.060	4,199	327	433	856
.060	[4,200]		[488]	
.060	4,204	565	711	1,295

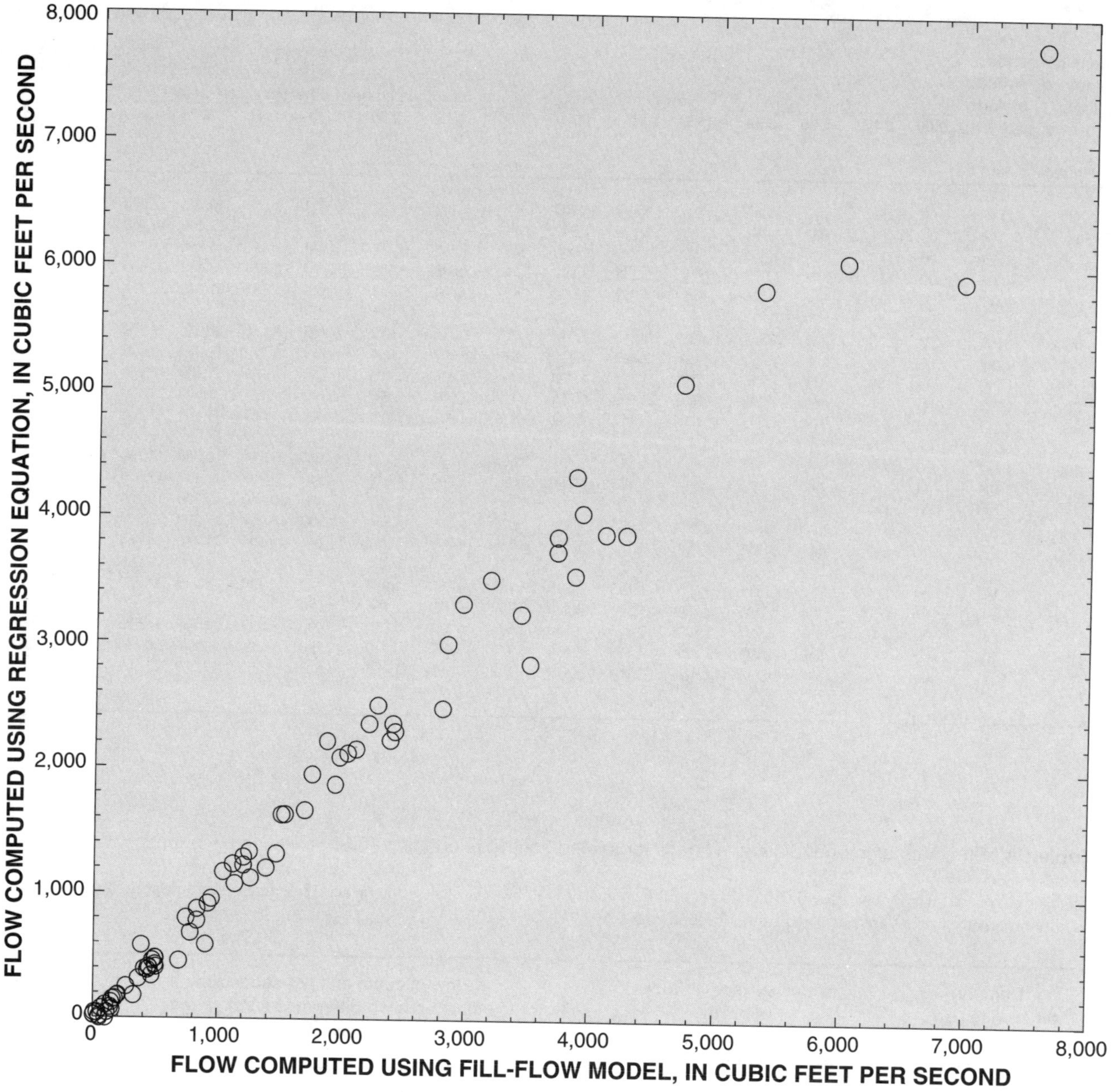

Figure C7. Relation of north-to-south flow (QNF) through the causeway fill of Great Salt Lake, Utah, computed using the regression equation to flow computed using the fill-flow model.

APPENDIX D. FLOW THROUGH THE CAUSEWAY CULVERTS

During 1957–59 when the causeway was constructed, the two 15-ft-wide culverts were placed where the lake is deepest. The bottoms of the culverts were about 10 ft above the lake bottom (Madison, 1970, p. 7 and 9). The bottom of the east culvert (fig. D1) was originally at an altitude of 4,180 ft and the bottom of the west culvert was at an altitude of 4,183 ft. The top of each culvert was originally at an altitude of 4,203 ft.

Two-way flow of brine can occur through the culverts (fig. D1) because of head and density differences between the south and north parts of the lake (ΔH and $\Delta\rho$). South-to-north flow (*QSC*, Appendix A, eq 3) through the upper part of the culverts is the result of a positive head difference (ΔH) at the causeway between the south and north parts. A positive density difference ($\Delta\rho$) between the north and south parts creates the potential for north-to-south flow (*QNC*, Appendix A, eq 4) through the lower part of the culverts.

The USGS measured the flow, *QSC* and *QNC*, through the east and west culverts using modified streamflow measurement techniques. The interface was located using velocity and density profiles.

Water-surface altitude, head difference, brine density, and computed and measured culvert flow during 1980–83 are listed in table D1. The culverts were not submerged from January 1980 to April 1983, and the culvert equations in the causeway model are valid for this period. The culverts were submerged from April 1983 through 1986 when the water surface of the south part rose above 4,203 ft. Because the culverts were probably filled with debris, measurement of culvert flow probably was not accurate. Some attempts were made to measure flow through the submerged culverts from April 1983 through 1986. Submerged conditions plus debris at the culvert openings made it difficult to make flow measurements at cross sections representative of the flow profile. The measurements made during the period were not included in table D1, and it was assumed for purposes of the study that the flow was small relative to that occurring through the fill and breach. There was probably no flow through the culverts from April 1983 through 1986.

In the original causeway model, Waddell and Bolke (1973) developed energy equations for estimating culvert flow, *QSC* and *QNC*. The head-loss coefficients were determined empirically from the energy equations by calibration with actual measurements. E.R. Holley (Holley and Waddell, 1976) developed an algorithm for a more rigorous treatment of the culvert flow (fig. D2). The equations of Holley and Waddell (1976) were used to replace the equations of Waddell and Bolke (1973). Waddell and Barton (1980) describe how the culvert equations were updated in this synopsis:

> Problems may occur when extrapolating empirical relations for smaller head and density differences than those actually observed; therefore, a mathematical model for the culvert flow was developed by E.R. Holley (Holley and Waddell, 1976). Holley's algorithm was used to determine if culvert flow computed by Waddell and Bolke (1973) was valid for the ranges of head and density differences that occurred during the

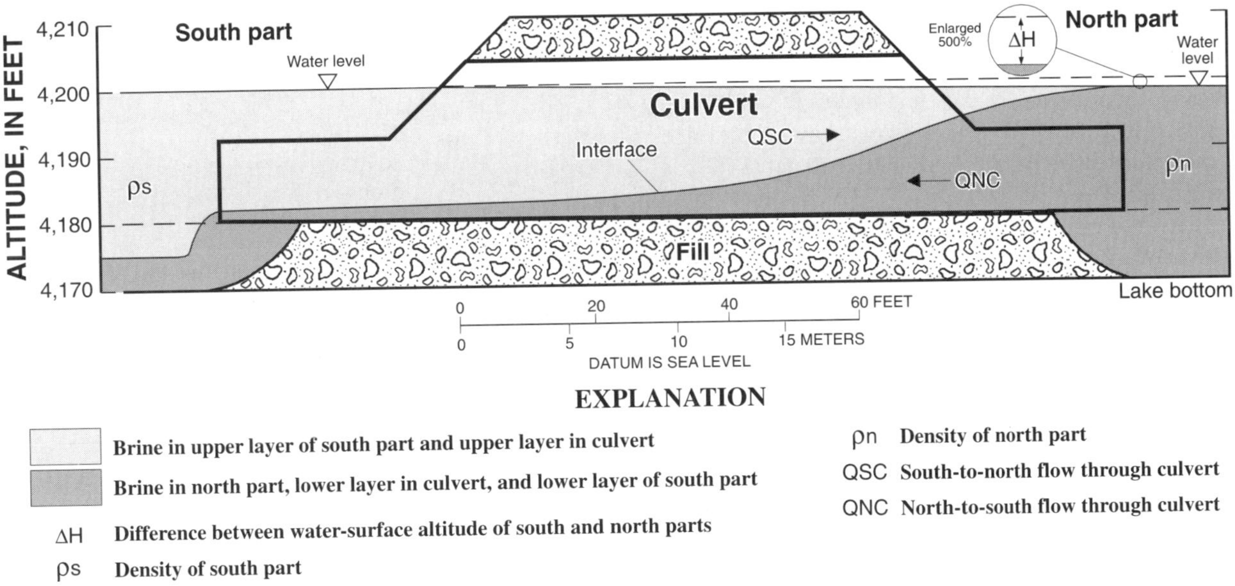

Figure D1. Schematic cross section of the east culvert in the causeway across Great Salt Lake, Utah, showing stratified flow and related hydraulic properties.

simulated period, for varying culvert widths. Flow determined by the two methods compared within standard error of about 20 percent.

The equations of Holley and Waddell (1976) have the capability of producing longitudinal profiles of the upper and interfacial surfaces as well as flow for varying head and density differences for three different flow regimes (fig. D2): (1) single layer with only brine from the south part in the culvert; (2) arrested wedge with only flow from south to north, but with a wedge of north-part brine intruding into the culvert and underlying the less dense brine from the south part; and (3) two-layer regime with the upper layer flowing from south to north and a lower layer flowing in the opposite direction (Waddell and Barton, 1980).

To compute *QSC* and *QNC* using the equations of Holley and Waddell (1976), the density of brine near the culverts was used to approximate the density of brine moving from north to south through the culverts. A method for estimating the density of the lower-layer brine as a function of ρn and *QSC* is described in Appendix E.

The altitude of the culvert bottom is important in determining *QNC*. The original altitudes of the culvert bottoms were known in 1959, but the culverts are often partially or completely filled with debris, which raises the effective bottom of the culverts. Measured culvert bottoms that give the effective bottom altitude of the culverts during the flow measurements are listed in table D1. Uncertainty in the altitude of the culvert bottom as a result of storms filling the culverts with debris between culvert flow measurements introduces some error in assuming that the average measured culvert bottom is representative of the average culvert bottom. To improve the relation between computed and measured culvert flow (in both directions, *QSC* and *QNC*), the altitude of the average culvert bottom was changed during model calibration to be 4,182 ft during 1980–83.

Correspondence between computed and measured values of *QSC* and *QNC* are shown in figures D3 and D4, respectively. The standard error of estimate as a percentage of the mean for computed and measured *QSC* and *QNC* was 12 percent and 62 percent, respectively. The relatively large standard error of estimate as a percentage of the mean for computed and measured *QNC* may be a result of uneven culvert bottoms created by debris in the culverts during measurement of flow.

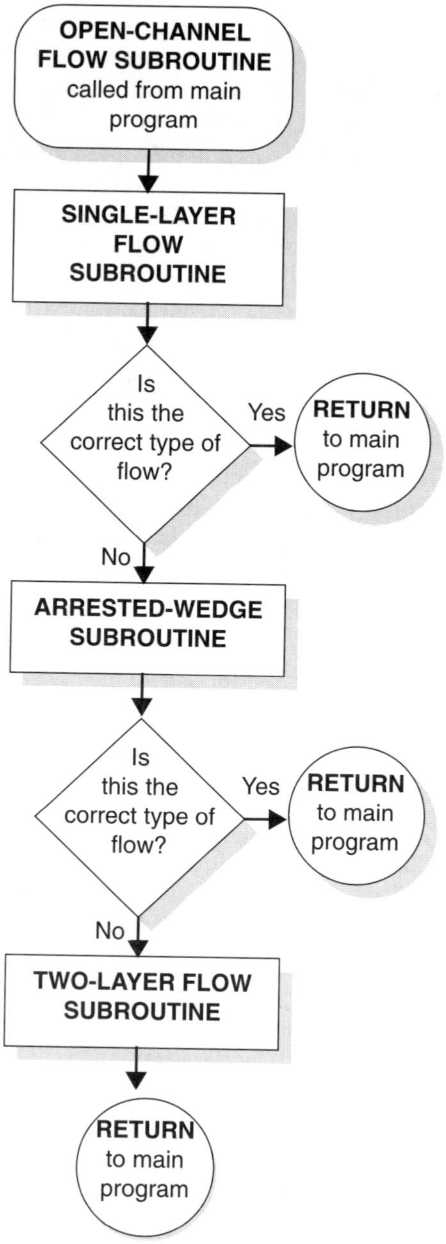

Figure D2. Flow chart of subroutine used to compute open-channel flow through the culverts and breach in the causeway across Great Salt Lake, Utah.

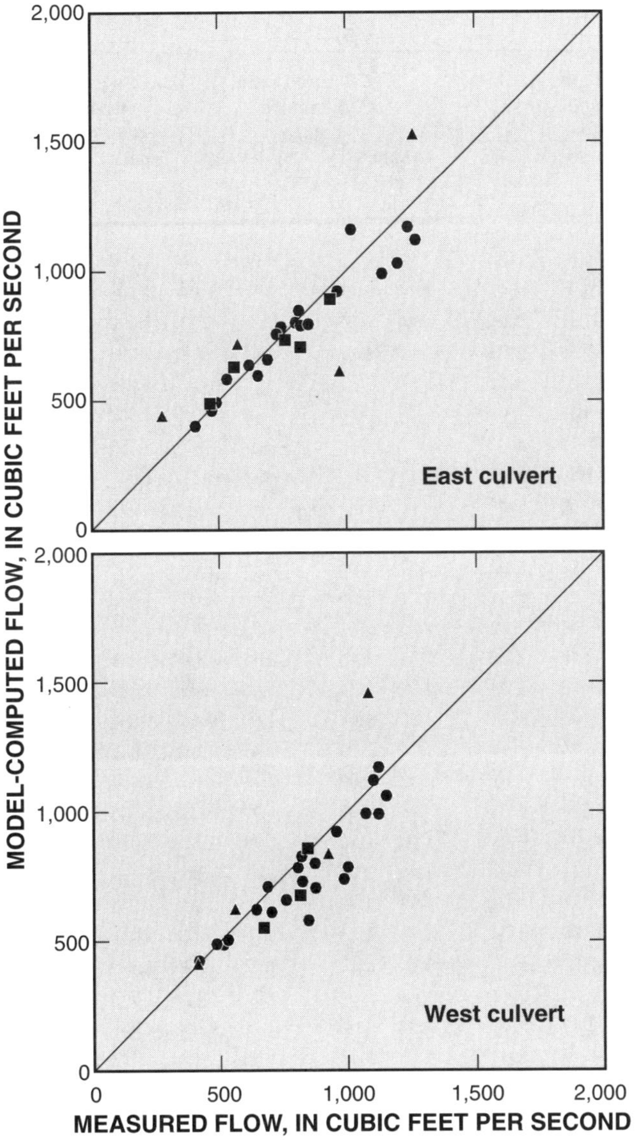

EXPLANATION

- ● **Steady conditions**—Head difference (ΔH) measured to less than 0.1 foot tolerance
- ■ **Unsteady conditions**—Head difference (ΔH) measured to greater than or equal to 0.1 foot tolerance
- ▲ **Windy conditions**—Indicated in table D1 as observed

Figure D3. Relation of model-computed to measured south-to-north flow (QSC) through the culverts in the causeway across Great Salt Lake, Utah, 1980–83.

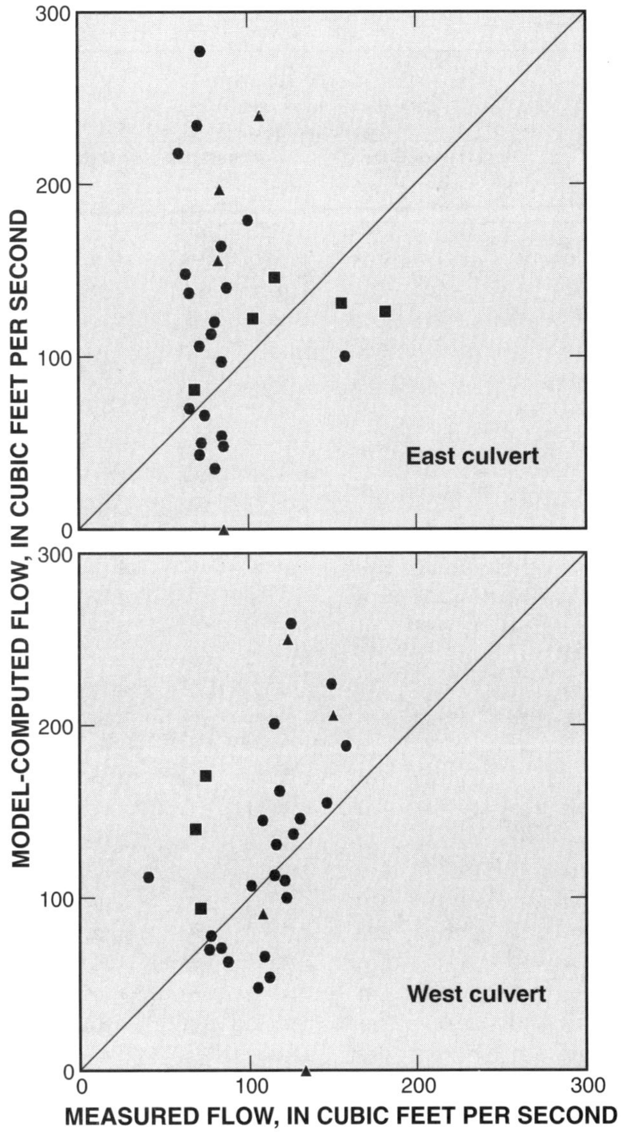

EXPLANATION

- ● **Steady conditions**—Head difference (ΔH) measured to less than 0.1 foot tolerance
- ■ **Unsteady conditions**—Head difference (ΔH) measured to greater than or equal to 0.1 foot tolerance
- ▲ **Windy conditions**—Indicated in table D1 as observed

Figure D4. Relation of model-computed to measured north-to-south flow (QNC) through the culverts in the causeway across Great Salt Lake, Utah, 1980–83.

Table D1. Water-surface altitude, head difference, and brine density in Great Salt Lake and flow through the east and

[Data from U.S. Geological Survey (1980, 1981) and ReMillard and others (1982, 1983). The terms ES, water-surface altitude of south part; QSC, south-to-north flow through culvert; and QNC, north-to-south flow through culvert are the same terms as those used in functioning; Estimated head difference at culvert: Difference between water-surface altitude of south and north parts minus 0.2 foot; Altitude of culvert bottom: Estimated from [(ES+EN-0.2)/2] - (water depth at south end of culvert); Flow through culvert: Type: T, two

Date	Water-surface altitude at gage during measurement of east culvert flow: Boat Harbor Gage, ES (ft)	Water-surface altitude at gage during measurement of east culvert flow: Saline Gage, EN (ft)	Water-surface altitude at gage during measurement of west culvert flow: Boat Harbor Gage, ES (ft)	Water-surface altitude at gage during measurement of west culvert flow: Saline Gage, EN (ft)	Estimated head difference at culvert: East ΔH (ft)	Estimated head difference at culvert: West ΔH (ft)	Average brine density: South part, ρs (g/mL)	Average brine density: North part, ρn (g/mL)
01-15-80	4,197.89 ±0.03	4,196.69 ±0.14	4,197.83 ±0.02	4,196.63 ±0.04	1.00 ±0.17	1.00 ±0.06	1.104	1.214
02-14-80	4,198.32 ±.01	4,196.98 ±.03	4,198.29 ±.01	4,196.96 ±.03	1.14 ±.04	1.13 ±.04	1.104	1.214
03-17-80	4,199.14 ±.02	4,197.48 ±.01	4,199.08 ±.02	4,197.51 ±.01	1.46 ±.03	1.37 ±.03	1.098	1.212
04-15-80	4,199.30 ±.04	4,197.54 ±.03	4,199.24 ±.06	4,197.58 ±.01	1.56 ±.07	1.46 ±.07	1.093	1.210
05-15-80	— —	4,197.94 ±.01	— —	4,198.02 ±.02	— —	— —	1.088	1.208
06-16-80	— —	4,198.27 ±.01	— —	4,198.26 ±.01	— —	— —	1.082	1.206
07-15-80	— —	4,198.22 ±.05	— —	4,198.15 ±.02	— —	— —	1.083	1.208
08-28-80	— —	4,197.81 ±.02	— —	4,197.80 ±.02	— —	— —	1.086	1.212
09-15-80	— —	4,197.68 ±.03	— —	4,197.69 ±.04	— —	— —	1.087	1.213
10-28-80	4,199.00 ±.03	4,197.53 ±.02	4,199.02 ±.02	4,197.53 ±.01	1.27 ±.05	1.29 ±.03	1.090	1.216
11-17-80	4,199.10 ±.04	4,197.53 ±.02	4,199.10 ±.03	4,197.54 ±.02	1.37 ±.06	1.36 ±.05	1.089	1.215
12-15-80	4,199.37 ±.03	4,197.66 ±.02	4,199.33 ±.03	4,197.67 ±.01	1.51 ±.05	1.46 ±.04	1.088	1.215
01-15-81	4,199.68 ±.01	4,197.92 ±.01	4,199.67 ±.01	4,197.92 ±.02	1.56 ±.02	1.55 ±.03	1.086	1.214
02-17-81	4,199.79 ±.12	4,198.11 ±.04	4,199.83 ±.06	4,198.18 ±.02	1.48 ±.16	1.45 ±.08	1.084	1.212
[1]04-15-81	4,199.76 ±.08	4,198.23 ±.06	4,200.02 ±.06	4,198.16 ±.03	1.33 ±.14	1.66 ±.09	1.081	1.211
06-15-81	4,20.13 ±.03	4,198.34 ±.01	4,200.12 ±.02	4,198.32 ±.03	1.59 ±.04	1.60 ±.05	1.080	1.210
07-15-81	4,199.59 ±.02	4,198.05 ±.01	4,199.57 ±.02	4,198.04 ±.01	1.34 ±.03	1.33 ±.03	1.082	1.211
08-17-81	4,198.86 ±.02	4,197.54 ±.01	4,198.88 ±.02	4,197.55 ±.03	1.12 ±.03	1.13 ±.05	1.084	1.212
09-14-81	4,198.46 ±.04	4,197.31 ±.03	4,198.43 ±.01	4,197.24 ±.02	.95 ±.07	.99 ±.03	1.086	1.213
[1]10-15-81	4,198.38 ±.04	4,197.18 ±.06	4,198.27 ±.02	4,197.09 ±.03	1.00 ±.10	.98 ±.05	1.088	1.213
11-16-81	4,198.37 ±.02	4,197.12 ±.02	4,198.41 ±.01	4,197.14 ±.01	1.05 ±.04	1.07 ±.02	1.090	1.213
02-16-82	4,198.91 ±.07	4,197.45 ±.03	4,198.89 ±.06	4,197.55 ±.08	1.26 ±.10	1.14 ±.14	1.095	1.212
03-15-82	4,199.48 ±.22	4,197.86 ±.06	4,199.46 ±.16	4,197.88 ±.04	1.42 ±.28	1.38 ±.20	1.087	1.212
04-15-82	4,20.18 ±.04	4,198.22 ±.08	4,200.16 ±.04	4,198.24 ±.10	1.76 ±.12	1.72 ±.14	1.078	1.211
05-14-82	4,20.57 ±.01	4,198.40 ±.03	4,200.59 ±.01	4,198.42 ±.01	1.97 ±.04	1.97 ±.02	1.069	1.211
06-15-82	4,20.70 ±.05	4,198.46 ±.01	4,200.70 ±.04	4,198.49 ±.01	2.04 ±.06	2.01 ±.05	1.066	1.211
08-16-82	4,199.95 ±.03	4,198.19 ±.02	4,199.89 ±.03	4,198.19 ±.02	1.56 ±.05	1.50 ±.05	1.077	1.212
[1]09-15-82	4,199.76 ±.03	4,198.04 ±.04	4,199.64 ±.04	4,198.07 ±.03	1.52 ±.07	1.37 ±.07	1.077	1.214
10-14-82	4,20.45 ±.01	4,198.61 ±.02	4,200.47 ±.01	4,198.57 ±.01	1.64 ±.03	1.70 ±.02	1.077	1.215
11-15-82	4,20.86 ±.02	4,198.82 ±.02	4,200.85 ±.02	4,198.78 ±.01	1.84 ±.04	1.87 ±.03	1.077	1.217
12-15-82	4,201.53 ±.05	4,199.20 ±.02	4,201.43 ±.03	4,199.22 ±.02	2.13 ±.07	2.01 ±.05	1.076	1.212
01-14-83	4,201.79 ±.01	4,199.54 ±.01	4,201.76 ±.02	4,199.53 ±.02	2.05 ±.02	2.03 ±.04	1.075	1.207
02-15-83	4,202.32 ±.01	4,20.05 ±.02	4,202.32 ±.02	4,20.05 ±.01	2.07 ±.03	2.07 ±.03	1.074	1.202
[1]03-14-83	4,202.81 ±.06	4,199.97 ±.03	4,202.80 ±.06	4,200.05 ±.10	2.64 ±.09	2.55 ±.16	1.071	1.199

[1]Unsteady conditions in culverts as a result of wind and waves.

west culverts in the causeway across Great Salt Lake, Utah, 1980–83

part; EN, water-surface altitude of north part; ΔH, head difference; ρs, average brine density of south part; ρn, average brine density of north equations and text throughout this report; ft, feet; g/mL, grams per milliliter; ft^3/s, cubic feet per second; ±, plus or minus; —, no data or gage not Estimated brine density north of culvert: Density in north part near culvert is reduced by mixing with brine from south-to-north flow; layer; A, arrested wedge]

Estimated brine density north of culvert		Water depth at south end of culvert		Altitude of culvert bottom		Flow through east culvert					Flow through west culvert				
						Measured		Computed			Measured		Computed		
East (g/mL)	West (g/mL)	East (ft)	West (ft)	East (ft)	West (ft)	QSC (ft^3/s)	QNC (ft^3/s)	QSC (ft^3/s)	QNC (ft^3/s)	Type	QSC (ft^3/s)	QNC (ft^3/s)	QSC (ft^3/s)	QNC (ft^3/s)	Type
1.212	1.212	16.2	17.8	4,181.0	4,179.3	466	116	487	197	T	513	108	505	271	T
1.212	1.211	16.3	18.5	4,181.3	4,179.0	533	78	568	150	T	846	40	593	245	T
1.209	1.208	17.8	20.0	4,180.4	4,178.2	803	72	787	113	T	984	83	743	240	T
1.207	1.206	18.2	18.6	4,180.1	4,179.7	816	71	859	107	T	1,000	87	778	159	T
1.204	1.204	19.2	20.8	—	—	983	74	—	—	—	1,130	71	—	—	—
1.201	1.201	21.0	22.5	—	—	1,310	52	—	—	—	1,390	92	—	—	—
1.205	1.204	19.8	21.0	—	—	1,010	69	—	—	—	1,140	114	—	—	—
1.210	1.209	17.0	19.8	—	—	596	83	—	—	—	663	201	—	—	—
1.211	1.211	16.9	20.6	—	—	621	75	—	—	—	656	217	—	—	—
1.213	1.213	19.7	19.6	4,178.5	4,178.6	655	84	619	345	T	702	146	649	321	T
1.213	1.213	19.5	19.6	4,178.7	4,178.6	694	65	683	296	T	759	126	689	299	T
1.212	1.212	20.0	19.9	4,178.4	4,178.5	741	84	785	259	T	822	121	740	281	T
1.211	1.211	21.0	20.4	4,177.7	4,178.3	747	158	805	306	T	805	122	810	271	T
1.210	1.209	21.5	20.8	4,177.4	4,178.1	762	156	773	364	T	873	130	736	342	T
1.207	1.207	22.2	20.2	4,176.7	4,178.8	973	83	670	489	T	923	108	881	224	T
1.207	1.206	20.5	21.0	4,178.6	4,178.1	823	80	825	277	T	871	115	852	290	T
1.209	1.209	20.2	20.2	4,178.5	4,178.5	620	100	655	371	T	641	158	655	371	T
1.211	1.210	19.2	14.0	4,178.9	4,184.1	492	70	525	401	T	530	149	488	130	T
1.211	1.211	18.4	13.0	4,179.4	4,184.7	408	72	417	430	T	415	125	396	136	T
1.212	1.211	17.0	14.7	4,180.7	4,182.9	277	107	446	313	T	409	123	424	196	T
1.211	1.211	18.4	13.4	4,179.2	4,184.3	474	59	499	355	T	484	115	468	105	T
1.210	1.210	19.7	13.0	4,178.4	4,185.1	561	182	653	296	T	670	74	521	57	T
1.209	1.209	20.5	14.0	4,178.1	4,184.6	821	103	729	316	T	813	68	667	41	T
1.208	1.208	21.2	15.0	4,177.9	4,184.1	937	68	916	272	T	842	71	844	26	T
1.207	1.207	20.5	17.0	4,178.9	4,182.4	1,140	65	1,020	200	T	1,070	76	993	56	T
1.207	1.207	21.5	20.0	4,178.0	4,179.5	1,200	74	1,060	245	T	1,120	77	1,010	187	T
1.209	1.210	21.0	19.6	4,178.0	4,179.3	729	87	783	348	T	686	118	731	300	T
1.212	1.212	20.8	20.1	4,178.0	4,178.7	575	82	740	376	T	557	150	646	394	T
1.212	1.212	21.0	19.6	4,178.4	4,179.8	853	63	834	329	T	819	116	848	234	T
1.213	1.214	21.5	20.0	4,178.2	4,179.7	966	71	954	290	T	955	101	959	203	T
1.208	1.208	21.3	20.5	4,179.0	4,179.7	1,020	80	1,170	152	T	1,150	109	1,090	154	T
1.203	1.203	19.0	20.8	4,181.6	4,179.7	1,270	84	1,130	62	T	1,100	112	1,130	141	T
1.198	1.198	19.6	21.0	4,181.5	4,180.1	1,240	85	1,180	64	T	1,120	105	1,190	114	T
1.195	1.196	21.0	19.5	4,180.3	4,181.8	1,260	85	1,570	0	A	1,080	133	1,470	0	A

APPENDIX E. FLOW THROUGH THE CAUSEWAY BREACH

A 300-ft-wide breach near the west side of Great Salt Lake (fig. 2) was opened August 1, 1984. The bottom of the breach was at an altitude of about 4,199.5 ft during 1984–86.

Two-way flow of brine can occur through the breach because of head and density differences between the south and north parts of the lake (ΔH and $\Delta \rho$). South-to-north flow (*QSB*, Appendix A, eq A3) through the upper part of the breach is the result of a positive head difference (ΔH) at the causeway between the south and north parts. A positive density difference ($\Delta \rho$) between the north and south parts creates the potential for north-to-south flow (*QNB*, Appendix A, eq A4) through the lower part of the breach.

The USGS measured the flow through the breach, *QSB* and *QNB*, using modified streamflow measurement techniques. Levels for altitude control were run, and the head difference at the causeway was determined between the south and north parts of the lake. The bottom profile within the breach opening was determined as part of the flow measurement. Brine samples were collected at selected depths from within the breach opening.

For several months after the breach was opened on August 1, 1984, brine flowed only to the north part; thereafter, two-layer flow generally was observed (table E1). The hydraulics governing flow through the breach and culverts is similar, so the equations of Holley and Waddell (1976) were used in the causeway model to compute breach flow (Appendix D).

To compute *QSB* and *QNB* using the equations of Holley and Waddell (1976), the density of the brine near the breach is used to approximate the density of brine moving from north to south through the breach. As brine from the south part enters the breach, it forms the upper layer of brine moving through the breach, which after exiting the breach, spreads out over the surface of the north part and mixes with the brine from the north part. This mixing causes the brine moving from north to south through the breach to be slightly lower in density than the average of the brine from the north part. Results of density measurements made about 400 ft north of the breach (fig. E1, site 1N) indicate that the effect of mixing is related to the amount of south-to-north flow (table E1). The density of the brine used as a boundary condition for computing flow through the breach is computed from equation E1.

$$\text{Density of brine near breach} = \rho n[1-(3x10^{-6})QSB] \quad \text{(E1)}$$

where

ρn = average density of brine in north part, in grams per milliliter, and

QSB = south-to-north flow through breach from previous time step, in cubic feet per second.

The cross-sectional opening of the breach is a trapezoidal shape with the east and west sides of the breach sloping into the lake. The equations of Holley and Waddell (1976) assume that openings in the causeway have vertical walls. An approximation was made to compute the cross-sectional area of the trapezoid formed by the sides of the breach, the water-surface altitude, and the bottom of the breach (table E1). The cross-sectional area was then used in conjunction with the equations of Holley and Waddell (1976) to compute breach flow.

Correspondence between the computed and measured values of *QSB* and *QNB*, respectively, is shown in figures E2 and E3. The standard error of estimate as a percentage of the mean for computed and measured *QSB* and *QNB* was 16 and 86 percent, respectively. The relatively large standard error of estimate as a percentage of the mean for computed and measured *QNB* is probably a result of the combination of measurement and model-computation errors caused by an uneven breach bottom created by debris in the breach during measurement of flow, and/or inaccurate representation of the cross section in the model.

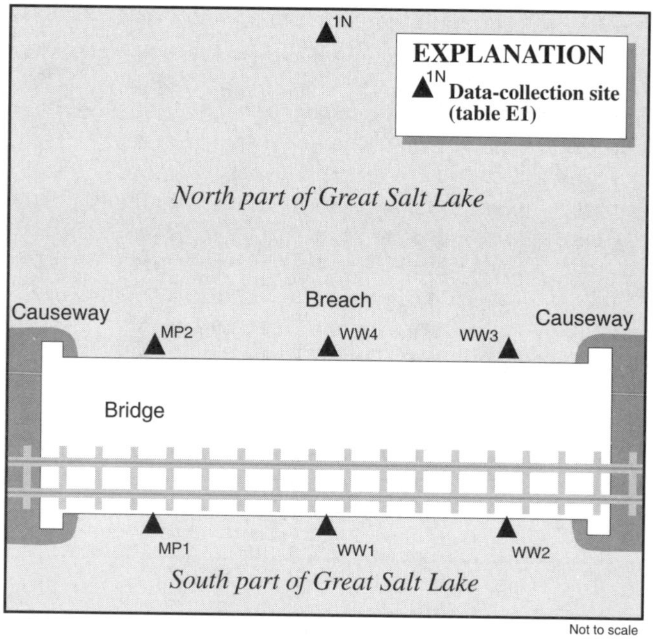

Figure E1. Schematic diagram showing approximate location of data-collection sites on and near the breach in the causeway across Great Salt Lake, Utah, 1984–86.

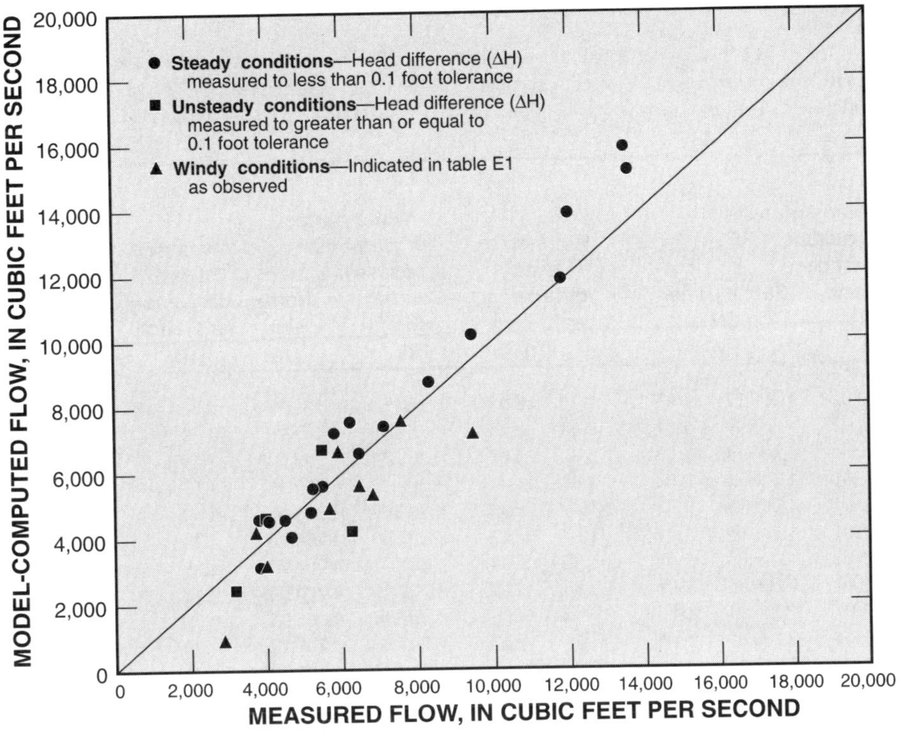

Figure E2. Relation of model-computed to measured south-to-north flow (QSB) through the breach in the causeway across Great Salt Lake, Utah, 1984–86.

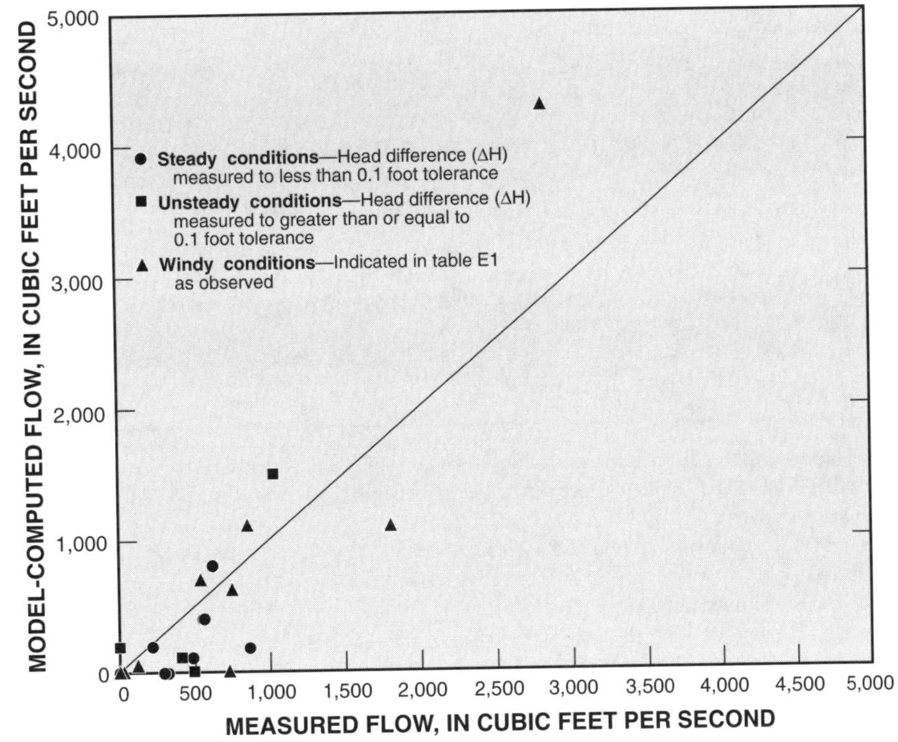

Figure E3. Relation of model-computed to measured north-to-south flow (QNB) through the breach in the causeway across Great Salt Lake, Utah, 1984–86.

Table E1. Water-surface altitude, head difference, and brine density in Great Salt Lake and flow through the breach

[See fig. E1 for site locations. The terms ES, water-surface altitude of south part; EN, water-surface altitude of north part; ΔH, head difference; flow through breach are the same terms as those used in equations and text throughout this report; ft, feet; g/mL, grams per milliliter; ft^3/s, cubic feet per foot; Average density north of breach: Average density above 4,200 feet at site 1N, $\rho 1N$; Breach width: Equivalent breach width averaged from width

Date	Water-surface altitude at gage: Boat Harbor Gage, ES (ft)		Water-surface altitude at gage: Saline Gage, EN (ft)		Estimated head difference at breach (ft)		Water-surface altitude at causeway staff: South side (ft)	Water-surface altitude at causeway staff: North side (ft)	Measured head difference at breach (ft)		Average density of south part, ρs (g/mL)	Average density north of breach, ρ1N (g/mL)
08-06-84	4,208.85	±0.02	4,205.72	±0.03	2.93	±0.05	[1]4,208.65	[2]4,205.77	2.84	—	[3]1.038	1.122
08-09-84	4,208.78	±.02	4,205.90	±.01	2.68	±.03	[1]4,208.58	[2]4,205.95	2.54	—	[3]1.038	1.120
08-14-84	4,208.44	±.04	4,206.17	±.03	2.07	±.07	[1]4,208.24	[2]4,206.22	2.19	—	[3]1.038	1.105
08-21-84	4,208.42	±.01	4,206.35	±.03	1.87	±.04	[1]4,208.22	[2]4,206.40	1.63	—	[3]1.039	[3]1.110
08-28-84	4,208.27	±.01	4,206.58	±.02	1.49	±.03	[1]4,208.07	[2]4,206.63	1.42	—	[3]1.039	1.124
09-04-84	4,208.10	±.01	4,206.65	±.01	1.25	±.02	[1]4,207.90	[2]4,206.70	1.20	—	[3]1.039	[3]1.123
09-13-84	4,207.97	±.02	4,206.85	±.01	.92	±.03	[1]4,207.77	[2]4,206.90	.99	—	1.039	1.121
09-26-84	4,207.90	±.01	4,206.85	±.01	.85	±.02	[1]4,207.70	[2]4,206.90	.80	±.01	[3]1.040	1.124
10-09-84	4,207.88	±.03	4,206.97	±.03	.71	±.06	4,207.72	4,206.95	.77	±.02	[3]1.041	1.121
10-30-84	4,208.01	±.01	4,207.05	±.06	.76	±.07	4,207.80	4,207.10	.70	±.05	[3]1.042	1.126
[4]11-27-84	4,208.28	±.02	4,207.33	—	.75	—	4,208.08	4,207.30	.78	±.04	[3]1.043	1.126
01-22-85	4,208.84	±.01	4,208.03	±.10	.61	±.11	4,208.55	4,207.95	.60	—	[3]1.043	1.117
03-12-85	4,209.47	±.07	4,208.52	±.20	.75	±.27	4,208.96	4,208.58	.38	±.23	1.043	[3]1.119
04-09-85	4,209.70	±.04	4,208.80	±.07	.70	±.11	4,209.43	4,208.86	.57	±.06	[3]1.043	[3]1.114
[4]05-07-85	4,209.80	±.12	4,208.44	±.02	1.16	±.14	4,209.75	4,208.89	.86	—	[3]1.043	[3]1.113
[4]07-02-85	4,209.54	±.01	4,208.88	±.01	.46	±.02	4,209.35	[5]4,208.89	.46	±.01	[3]1.044	[3]1.118
07-16-85	4,209.30	±.05	4,208.57	±.08	.53	±.13	4,209.22	4,208.64	.58	±.08	[3]1.045	1.121
08-20-85	4,208.76	±.02	4,208.07	±.08	.49	±.10	4,208.55	4,208.07	.48	±.03	[3]1.046	1.124
09-24-85	4,208.47	±.08	4,207.67	±.09	.60	±.17	4,208.16	4,207.60	.56	±.04	[3]1.047	1.123
[4]10-29-85	[6]4,208.36	—	[7]4,207.57	—	[8].59	—	4,208.16	4,207.62	.54	±.04	[3]1.048	[3]1.120
12-10-85	4,208.74	±.20	4,207.90	—	[8].64	—	4,208.70	4,208.18	.52	±.40	[3]1.049	[3]1.115
01-07-86	4,209.00	±.04	4,208.18	±.01	.62	±.05	4,208.71	4,208.15	.56	±.06	[3]1.050	[3]1.117
02-25-86	4,209.79	±.01	4,208.85	±.01	.74	±.02	[1]4,209.59	[2]4,208.90	.72	±.01	1.052	[3]1.104
04-01-86	4,210.50	±.10	4,209.55	±.10	.75	±.20	4,210.28	4,209.54	.74	±.09	[3]1.049	[3]1.105
[4]05-14-86	[9]4,211.75	±.05	4,210.43	±.08	1.12	±.13	[1]4,211.55	[2]4,210.48	.53	±.28	[3]1.046	[3]1.104
[4]05-27-86	[9]4,211.76	±.01	4,210.70	±.05	.86	±.06	[1]4,211.56	[2]4,210.75	.64	±.02	[3]1.045	[3]1.102
American Magnesium Corporation dike broke on June 8, 1986												
[4]06-10-86	[9]4,211.47	±.03	[10]4,211.09	—	[8].18	—	[1]4,211.27	[11]4,211.14	.13	±.55	1.043	[3]1.100
06-24-86	[9]4,211.53	±.02	4,210.60	±.01	.73	±.03	[1]4,211.33	[2]4,210.65	.64	±.03	[3]1.043	[3]1.102
[4]07-29-86	[9]4,211.15	±.12	[10]4,210.44	—	[8].51	—	[1]4,210.95	[11]4,210.49	.46	±.02	[3]1.044	1.106
09-23-86	[9]4,210.73	±.05	[10]4,209.79	—	[8].74	—	[1]4,210.53	[11]4,209.84	.69	±.14	[3]1.045	[3]1.105
[4]10-15-86	[9]4,210.85	±.01	4,208.95	±.02	1.70	±.03	[1]4,210.65	[2]4,209.00	[12].68	±.15	[3]1.045	[3]1.105
[4]12-02-86	[9]4,211.08	±.02	4,210.10	±.04	.78	±.06	[1]4,210.88	[2]4,210.15	.54	±.15	[3]1.046	[3]1.106

[1]Estimated from ES - 0.20 foot, which is the average difference between ES and water-surface altitude at south staff.
[2]Estimated from EN + 0.05 foot, which is the average difference between EN and water-surface altitude at north staff.
[3]Interpolated between measured data.
[4]Unsteady conditions in breach as a result of wind and waves.
[5]North is calm, south is rough.
[6]Estimated from water-surface altitude at south staff + 0.20 foot.
[7]Estimated from water-surface altitude at north staff - 0.05 foot.
[8]Estimated from ES - EN - 0.2 foot.
[9]Between 04-01-86 and 05-14-86, the Boat Harbor Gage developed a 0.5-foot error.
[10]Gage out, estimated from water-surface altitude at north staff - 0.05 foot.
[11]Estimated from water-surface altitude at south staff - ΔH.
[12]Revised to 0.86.

ρs, average density of south part; ρn, average density of north part; 1N, site 1N; QSB, south-to-north flow through breach; and QNB, north-to-south second; ±, plus or minus; —, no data; Estimated head difference at breach: Difference between water-surface altitude of south and north parts minus 0.2 at ES and breach bottom width of 215 feet; Breach flow: Type: S, single layer; A, arrested wedge; T, two layer]

			Breach flow					Water-surface altitude at breach					
			Measured		Computed			West end		Middle		East end	
Average density of north part, ρn (g/mL)	Ratio of ρ1N to ρn	Breach width (ft)	QSB (ft^3/s)	QNB (ft^3/s)	QSB (ft^3/s)	QNB (ft^3/s)	Type	South side, site MP1 (ft)	North side, site MP2 (ft)	South side, site WW1 (ft)	North side, site WW4 (ft)	South side, site WW2 (ft)	North side, site WW3 (ft)
[3]1.167	0.961	245	13,500	0	15,900	0	S	—	—	—	—	—	—
[3]1.167	.960	245	13,600	0	15,200	0	S	—	—	—	—	—	—
[3]1.166	.948	244	12,000	0	13,900	0	S	—	—	—	—	—	—
[3]1.164	.954	244	11,800	0	11,900	0	S	—	—	—	—	—	—
[3]1.163	.966	243	9,440	0	10,200	0	S	—	—	—	—	—	—
[3]1.162	.966	243	8,300	0	8,780	0	S	—	—	—	—	—	—
1.160	.966	242	5,800	0	7,240	0	A	—	—	—	—	—	—
[3]1.161	.968	242	5,490	0	5,630	0	A	—	—	—	—	—	—
[3]1.162	.965	242	5,230	0	5,560	0	A	4,207.50	4,207.50	4,207.52	4,207.40	4,207.66	4,207.36
[3]1.163	.968	242	5,180	0	4,840	0	A	4,207.53	4,207.40	4,207.50	4,207.47	4,207.70	4,207.48
[3]1.163	.968	243	6,450	0	5,640	0	A	4,208.00	4,207.86	4,208.85	4,207.76	4,208.02	4,207.76
[3]1.156	.966	245	3,970	489	4,670	119	T	4,208.33	4,208.35	4,208.32	4,208.26	4,208.48	4,208.28
1.150	.973	247	3,170	1,030	2,470	1,510	T	4,208.63	4,208.48	4,208.55	4,208.72	4,208.80	4,208.50
[3]1.147	.971	248	4,490	560	4,610	411	T	4,209.35	4,209.20	4,209.20	4,209.14	4,209.32	4,209.08
[3]1.145	.972	248	7,570	26	7,620	0	A	4,209.48	4,209.32	4,209.46	4,209.41	4,209.73	4,209.40
[3]1.142	.979	247	4,005	854	3,220	1,130	T	4,209.75	4,209.18	4,209.15	4,209.12	4,209.28	4,209.12
[3]1.142	.982	247	4,060	555	4,570	412	T	4,209.56	4,208.86	4,209.00	4,208.92	4,209.12	4,208.72
[3]1.142	.984	245	3,830	620	3,170	814	T	4,208.91	4,208.30	4,208.35	4,208.37	4,208.46	4,208.14
[3]1.142	.983	244	4,660	220	4,090	201	T	4,208.56	4,207.93	4,207.93	4,207.90	4,208.08	4,207.73
[3]1.142	.981	244	3,709	474	4,000	192	T	4,208.52	4,208.03	4,208.40	4,208.07	4,208.02	4,207.72
[3]1.142	.976	245	6,240	0	4,260	202	T	4,209.06	4,208.43	4,208.34	4,208.52	4,208.48	4,208.27
[3]1.142	.978	246	3,800	481	4,620	115	T	4,209.08	4,208.46	4,208.46	4,208.38	4,208.52	4,208.41
1.142	.967	248	7,110	296	7,440	0	A	—	—	—	—	—	—
[3]1.139	.970	250	6,224	322	7,580	0	A	4,210.60	4,210.00	4,210.05	4,209.92	4,210.08	4,209.70
[3]1.134	.974	254	6,818	538	5,370	721	T	—	—	—	—	—	—
[3]1.132	.973	254	9,455	120	7,210	72	T	—	—	—	—	—	—
American Magnesium Corporation dike broke on June 8, 1986													
[3]1.131	.973	253	2,860	2,820	947	4,310	T	—	—	—	—	—	—
—	—	254	6,450	864	6,620	187	T	—	—	—	—	—	—
—	—	252	3,720	1,800	4,240	1,120	T	—	—	—	—	—	—
—	—	251	5,470	489	6,740	0	A	—	—	—	—	—	—
—	—	252	5,900	723	6,680	0	A	—	—	—	—	—	—
—	—	252	5,670	745	4,950	644	T	—	—	—	—	—	—

APPENDIX F. INITIAL CONDITIONS AND INPUT USED FOR CALIBRATION OF CAUSEWAY MODEL

Calibration of the causeway model required the use of inflow data from 1980–86 as input to the causeway model. The compilation of these data are discussed in Appendix A. The physical and chemical conditions of the lake on January 1, 1980, were used as initial conditions to begin model computations for the 1980–86 calibration period:

Initial Lake Conditions

Dissolved salt load in south part	*LS*	1.80 billion tons
Dissolved salt load in deep brine layer of south part	*LSL*	0.30 billion tons
Dissolved salt load in north part	*LN*	2.13 billion tons
Cumulative precipitated salt load in north part	*CLNP*	0.67 billion tons
Water-surface altitude of south part	*ES*	4,197.70 ft
Difference between water-surface altitude of south and north part at the causeway (head difference)	ΔH	1.05 ft

Causeway Conditions

Breach width: variable from 215 to 255 ft, depending on water-surface altitude

Breach bottom	4,199.5 ft
Culvert top	4,203.0 ft
Culvert bottom	4,182.0 ft
Culvert width	30.0 ft

Calibration of the causeway model is discussed in the main body of this report. The model used a constant time interval, *T* equals 1.901 days, during calibration.

GLOSSARY

Abbreviation	Definition	Unit
AN	Area of north part	Acres
AS	Area of south part	Acres
CF	1/12 (conversion between inches and feet)	Foot per inch
CLNP	Cumulative precipitated salt load in north part	Tons
CM	Time interval (in this case, 365/12 days, about 1 month)	Days
CN	Dissolved-solids concentration in the north part	Grams per milliliter
CS	Dissolved-solids concentration in the south part	Grams per milliliter
EAAN	Average annual freshwater evaporation from north part	Inches
EAAS	Average annual freshwater evaporation from south part	Inches
EAI	Fraction of average annual evaporation	—
EMI	Fraction of average monthly evaporation	—
EN	Water-surface altitude of north part	Feet
EON	Evaporation from north part	Acre-feet per day
EOS	Evaporation from south part	Acre-feet per day
ES	Water-surface altitude of south part	Feet
GIN	Ground-water inflow to north part	Acre-feet per day
GIS	Ground-water inflow to south part	Acre-feet per day
ΔH	Difference between water-surface altitude of south and north part at the causeway (head difference)	Feet
I	Time step (number of elapsed time intervals)	—
IR	Ratio of inflow used for the 10-year simulated period to 1981 inflow	—
LN	Dissolved salt load in north part	Tons
LND	Redissolved salt in north part	Tons
LNP	Precipitated salt load in north part	Tons
LS	Dissolved salt load in south part	Tons
LSD	Redissolved salt in south part	Tons
LSL	Dissolved salt load in deep brine layer in south part	Tons
LSP	Precipitated salt load in south part	Tons
LT	Total salt load in lake	Tons
PIN	Precipitation in north part	Acre-feet per day
PIS	Precipitation in south part	Acre-feet per day
PAAN	Average annual precipitation (1931–73) for the north part of the lake	Inches
PAAOGSF	Average annual precipitation (1931–73) at Ogden Sugar Factory	Inches
PAAS	Average annual precipitation (1931–73) for the south part of the lake	Inches
PAASLC	Average annual precipitation (1931–73) at Salt Lake City Airport	Inches
PAATOOL	Average annual precipitation (1931–73) at Tooele	Inches
PMOGSF	Measured monthly precipitation at Ogden Sugar Factory	Inches

GLOSSARY—Continued

Abbreviation	Definition	Unit
PMSLC	Measured monthly precipitation at Salt Lake City Airport	Inches
PMTOOL	Measured monthly precipitation at Tooele	Inches
PRT	Average ratio index for precipitation	—
QN	Total north-to-south flow through causeway	Acre-feet per day
QNB	North-to-south flow through breach	Cubic feet per second
QNC	North-to-south flow through culvert	Cubic feet per second
QNF	North-to-south flow through fill	Cubic feet per second
QS	Total south-to-north flow through causeway	Acre-feet per day
QSB	South-to-north flow through breach	Cubic feet per second
QSC	South-to-north flow through culvert	Cubic feet per second
QSF	South-to-north flow through fill	Cubic feet per second
SCFN	Salinity correction factor for evaporation rate in the north part	—
SCFS	Salinity correction factor for evaporation rate in the south part	—
SIS	Surface-water inflow to south part	Acre-feet per day
VN	Volume of north part	Acre-feet
VS	Volume of south part	Acre-feet
ΔVN	Change in volume of north part	Acre-feet
ΔVS	Change in volume of south part	Acre-feet
T	Time interval (in this study, *T* equals 1.901)	Days
YNF	Average thickness of the lower layer of brine flowing from north to south through fill	Feet
YSF	Average thickness of the upper layer of brine flowing from south to north through fill	Feet
$\Delta\rho$	Density difference between brine in north and south part	Grams per milliliter
ρn	Density of brine in north part at any temperature	Grams per milliliter
ρs	Density of brine in south part at any temperature	Grams per milliliter

☆U.S. GOVERNMENT PRINTING OFFICE: 1997-646-425